AMICS

CTURE AND
OPERTIES

C. A. DANIELS, PhD, P.E.
Senior Materials Engineer
Affiliated Consulting Engineers

Abyss Books
Washington, D.C.
www.abyssbooks.ws

Published By:
    Abyss Books
    Post Office Box 4913
    Washington D.C. 20008 USA
    www.abyssbooks.ws

Copyright 2002 by C. A. Daniels

Design and illustrations by C. A. Daniels

ISBN 0-9713404-0-4

Printed and bound by BookMasters, Inc.
www.bookmasters.com

9 8 7 6 5 4 3 2 1

TABLE OF CONTENTS

Chapter 4, TENSILE PROPERTIES

Chapter 5, MECHANICAL PROPERTIES

DATA SECTION

BONDING AND STRUCTURE

BONDING TYPE

All forms of chemical bonding are represented within the huge diversity of ceramics. The most common type is the covalent bond modified by some level of ionic character. Only when a bond is between the same kind of atom, such as within a diamond, will the bond be purely covalent. The amount of deviation from pure covalent bonding that a specific bond displays may most easily be envisioned by considering the relative size of the atoms involved. The longer the distance that an electron travels in the orbit around an atom the more the atom displays the negative charge of that electron as compared to the other atom with which it is sharing the electron. This partial ionic character of most ceramic bonds affects many properties, especially electrical ones.

Figure 1-1 shows bonding shapes for three materials. The s-s bond has no directionality. The s-p and p-p bonds both involve p orbitals and are consequently directional. The three p orbitals of an atom are perpendicular to each other and allow little deviation from this angular relationship. Only when atoms are appropriately aligned can a directionally dependent bond be produced or maintained.

A purely ionic bond comes about when one atom removes an electron or electrons from another atom. Ions must be physically removed from each other to prevent recombination. This most often occurs when a partially ionic bonded ceramic is dissolved. Sodium chloride in water forms discreet ions of both sodium and chlorine.

Secondary bonding within a ceramic may be the primary means of forming a solid. For a molecule like water, bonding is fully satisfied within the molecule. In order to have some inter-molecule bonding the water molecules arrange themselves such that the larger oxygen atoms are nearer the smaller hydrogen atoms of the neighboring atom. Some ionic attraction can then be achieved to augment the weak mass attraction between low-density water molecules. In group, eight elements not even ionic attraction is operable. As a result, helium does not solidify because it has only very weak mass attraction available to it.

9

BONDING STRENGTH

An evaluation of bond length and bond type will give insight into bond strength and the resulting overall strength of the ceramic in the bond direction. Experiments to test individual bonds are fairly well developed. Using computer modeling, this information is processed to predict possible new ceramic materials with desirable properties. Short bond length and strong covalent bonding is an indicator of a strong ceramic, but the theory is not fully developed.

Although individual bond strength between two atoms is a consideration, the overall bonding strength of a ceramic is of more general concern. A ceramic of mixed strong and weak bonds is not as useful in most applications as a ceramic of uniformly moderate bonding strength. How the ceramic crystallizes effects the overall bonding strength.

CHEMICAL BONDING WITHIN CRYSTAL STRUCTURE

Elements singularly or in various combinations can be chemically reacted to form a three dimensionally bonded crystalline ceramic. Just which elements can combine and in what proportions to produce a stable desirable result is determined primarily by the three criteria:

First, the product must be electrically neutral on the macroscopic scale. If the total electronic charges on the component elements are zero then this is fulfilled even though localized charges may be unbalanced.

Second, the component elements must be able to fit together physically so that electron sharing can take place. This is a particular problem when very small atoms and very large atoms are brought together. It can always be accomplished for a single bond in one direction to form a molecule, but this will not suffice if the result is to be used in the solid state. Additionally, any angularity of the bond must be accommodated.

Third, the individual elements of the material must be able to be arranged in a repeatable pattern in all directions if the product is to crystallize. Although these three restrictions make the possible solid state ceramics considerably less in number than would otherwise be imaginable, the variety and complexity of crystalline ceramics is still truly amazing.

When chemical bonding within a crystal is three dimensional,

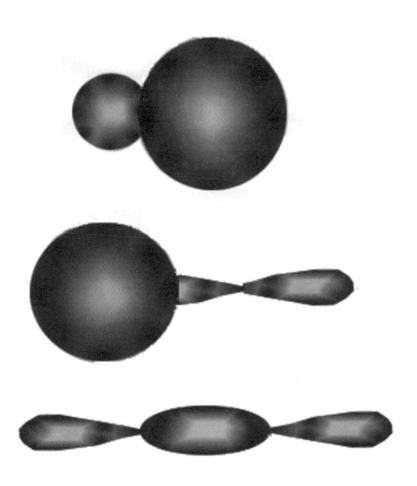

Figure 1-1 THREE CONFIGURATIONS OF CHEMICAL BONDS
The top diagram shows the relative sizes of sodium and chloride in solid state bonding, both of which have s orbital bonding. The electrons spend more time orbiting the chloride and thus render it more negative in character than the sodium. The chlorine and the sodium may be in any orientation and maintain their bond. The middle diagram is of a mixed s and p orbital such as diamond displays. In this case the p orbital determines the direction of the bond. The bottom diagram is of two p orbitals which graphite displays in its c axis. The double p bond is not only highly directional, it is also very long in physical distance. Consequently, the c axis graphite bond is weaker than the diamond bond.

an individual atom is bonded to all of its immediate neighbors. The bondings between nearest neighbors are not necessarily equal in strength, type and length. The electrons associated with a particular atom are free to move throughout the larger crystal provided there is a compensating drift of electrons to maintain electrical neutrality. Thus, atoms are not exclusively bound to their neighbors; they are effectively bound to the entire crystal.

CRYSTAL STRUCTURE OF CERAMICS

A crystal forms by filling space in a repeatable pattern. Mathematically this can be accomplished if the pattern to be repeated can be contained within one of 14 different shapes. These shapes, Bravais Lattices, are relatively simple in themselves but the number and arrangements of the atoms that can be contained within each one of these repeatable shapes, unit cells, can become very complex. Fortunately some of the most common ceramics are also the most simple both in the shape of the unit cell and in the number of atoms contained within the unit cell. These simple ceramics are often used to develop models of ceramic behavior and the results applied to the more complex ceramics.

Miller Indices were developed to describe planes within the unit cell. A schematic of a unit cell with the most common indices marked is shown in Figure 1-2. Figure 1-3 shows a simple cubic unit cell with the atoms indicated. The concepts of Figures 1-2 and 1-3 are combined in Figure 1-4. From the planes shown in these figures it is clear that atoms within a unit cell are more closely packed in some planes than in others. The relative tightness of atom packing or. atomic density within a plane is called the packing factor. Packing can also be considered along a specific line within a plane, the packing direction.

The simple cubic structure shown in Figures 1-2 through 1-4 is the least complicated structure. It is not an efficient one as can be inferred from all of the open space. Thus, although this is a possible structure it is not one seen in nature. The simplest naturally occurring structure is that of cesium chloride, which has a simple cubic structure but has two atoms occupying a lattice site instead of one. The resulting structure does not really look like the simple cubic structure shown in Figure 1-4 but more like a body centered cubic structure. Figure 1.5 shows the cesium chloride structure, which can be compared to the body centered cubic structure shown in Figure 1-6. The remaining

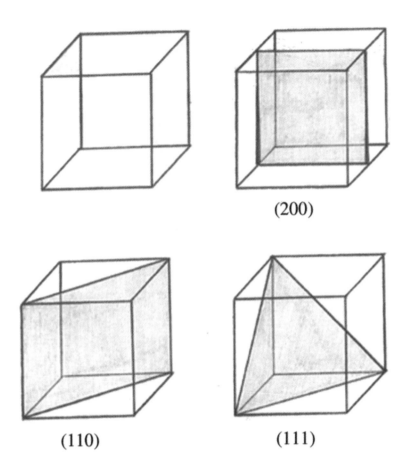

(200)

(110)   (111)

Figure 1-2 SCHEMATIC OF A UNIT CELL
A unit cell is the smallest repeatable entity that characterizes the crystalline structure. The Miller Indies noted are the most commonly used planes when discussing
a ceramic on the atomic or unit cell level. All of the surfaces of the cube shown at
the upper left are (100) type planes and are interchangeable. If the unit cell is not a
cube than the distinction between (100), (010) or (001) planes must be made. The
plane that slices the unit cell in half and is parallel to the (100) plane is the (200)
plane, which is shown as the shaded area in the upper right cube. The shaded area
within the cube on the lower left is the (110) plane. The plane that slices through the
unit cell on the diagonal is the (111) plane marked by the shaded area within the
lower right cube.

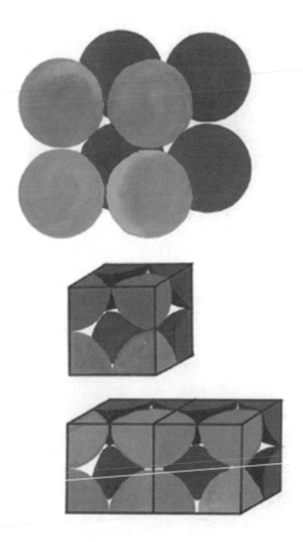

Figure 1-3 DIAGRAM OF THE UNIT CELL FOR SIMPLE CUBIC
The most simple unit cell is a cubic shape with an atom in each corner. To fill each
corner requires eight atoms and looks like the sketch on the top. The lighter color
indicates closer atoms and the darker color indicates further atoms. To draw a unit
cell for this arrangement the construction lines originate from the centers of each of
the atoms at the corners. This results in the construction of the unit cell slicing the
corner atoms up into eighths. Since each atom is part of the unit cell shown plus the
others that can be drawn throughout space the unit cell drawn only contains one
eighth of the shown corner atoms. The other seven pieces of each atom are in the
seven surrounding unit cells. The middle figure shows the segment of the atoms
associated with just this unit cell and the bottom figure shows two unit cells together

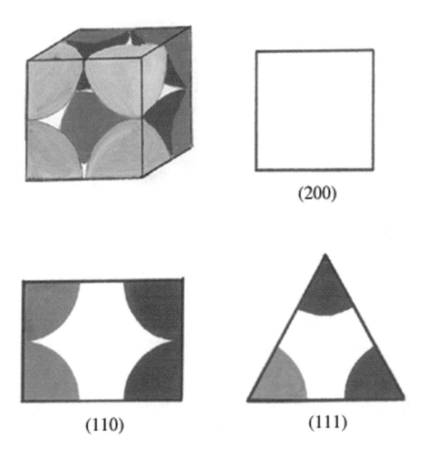

(200)

(110)   (111)

Figure 1-4 SCHEMATIC OF SIMPLE UNIT CELL WITH PLANES SHOWN
SEPARATELY
From the unit cell diagram in Figure 1-3 it is clear that drawing the planes shown in
Figure 1-2 will reveal the spatial relationships between atoms in each plane. The
upper left diagram is a repeat of the middle diagram in Figure 1-3 showing the (100)
planes. The top right diagram shows the (200) plane. This is completely blank
because, this plane slices none of the atoms in this unit cell. The lower left diagram
shows the (110) plane redrawn flat relative to the page. Note the greater spacing of
the atoms in the long direction. The bottom right diagram depicts the (111) plane.
The spacing between atoms along the sides of the triangle is the same as the long
direction of the (110) plane because the distance and direction are the same. The
center is completely empty, as is the center of the unit cell through which this plane
slices. Making these slices through a unit cell helps to visualize the effect that directionality has on the properties of the crystal. The strength in the planes with the
more open space will be less than the planes more completely occupied with atoms,
provided the bonding is uniform.

15

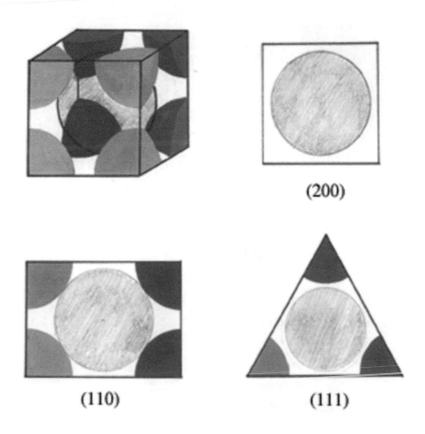

(200)

(110)   (111)

Figure 1-5 SCHEMATIC OF CESIUM CHLORIDE STRUCTURE

The above are the diagrams for cesium chloride. It has a simple cubic structure with a cesium and a chloride atom associated with each site of the unit cell. The second atom is situated relative to the corner such that it slips into the center. Here the corner atoms are the cesium, shown as solid circles, and the shaded circles correspond to the chloride. From the planes based on Figure 1-4, it is clear that this structure, with a pair of atoms per site, looks quite different from a single atom associated with a site.

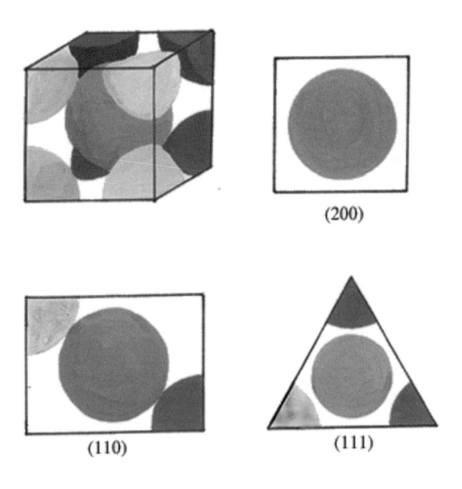

(200)

(110)

(111)

Figure 1-6 SCHEMATIC OF THE BODY CENTERED CUBIC STRUCTURE
Arranged as in Figure 1-4, these diagrams are for a simple body centered cubic crystal where the atoms are all the same type. Note the similarity between this arrangement and that of cesium chloride in Figure 1-5.

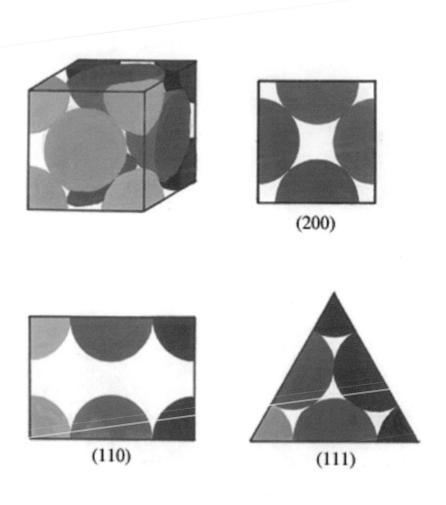

(200)

(110)

(111)

Figure 1-7 SCHEMATIC OF FACE CENTERED CUBIC STRUCTURE
As in Figure 1-4, the diagrams are arranged such that the unit cell and three planes though this cell are shown. Like the simple cubic and the body-centered cubic, this unit cell is also a cube. As the name implies, atoms are located on the face of each side of the cube. Adding atoms here produces a very densely packed plane and makes the over all density of this structure very high. Usually ceramics in this structure are at their most dense.

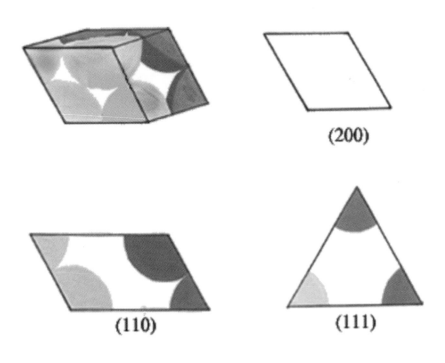

Figure 1-8 SCHEMATICS OF THE RHOMBOHEDRAL STRUCTURE
This figure shows diagrams for the rhombohedral structure with the same planes as shown in Figure 1-4. The rhombohedron has sides of the same length but is without right angle corners. Note that the (110) plane in different directions are not the same area or orientation.

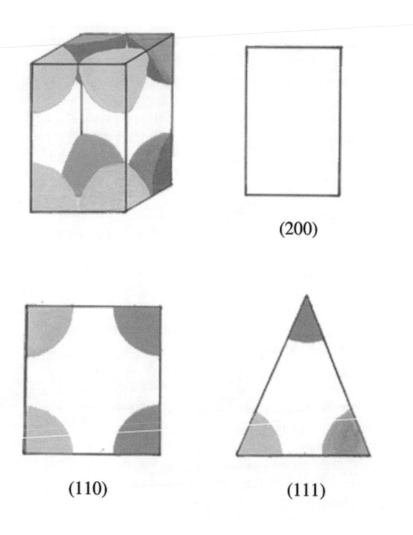

(200)

(110)   (111)

Figure 1-9 SCHEMATIC OF THE TETRAGONAL STRUCTURE
This figure shows diagrams, in the Figure 1-4 format, for the tetragonal structure.
The planes of the unit cell are at right angles but the lengths are not all equal. The
top plane forms a square and the side faces are rectangles. This structure also has a
body centered version where another atom is located in the center of the unit cell.

possible crystalline types are shown in Figures 1-7 through 1-12.

CRYSTALLIZATION

If crystallization is begun from the liquid state, the process of nucleation and growth creates the crystals. Nucleation is the process wherein the solid state has reached a sufficient level of mass that growth will continue at that site. Often within the melt random motion will bring together aligned atoms or molecules such that a few unit cells of the crystal may actually form, but this will not continue to grow if it is not of sufficient size. It is often an alien site such as an impurity or a deliberately added seed crystal, which will form the rallying point for a critical number of unit cells to amass and become a nucleus.

Once the nucleus is established the crystal will continue to grow as available aligned material comes within proximity. The usual direction of quick growth is in a loosely packed plane because less material needs to be available to that plane for growth as compared to a plane with a higher packing factor. Thus, if single crystal growth is allowed to proceed for a time and then stopped before other crystals impinge upon it, the final single crystal will have a characteristic shape with well defined growth planes. There are many beautiful examples of this among the natural gemstones.

SINGLE CRYSTAL STRUCTURES

Plate like single crystal structures of rock surrounded by amorphous material comprise clay. These structures did not necessarily form in this shape. The weathering action of water, sun and wind reverse the usual pattern of fast growing, low density planes being on the surface. Erosion preferentially removes the more loosely bonded atoms leaving the surface with the slow growing, high-density planes exposed.

Some types of clay are layers of ceramic resembling a sandwich. The sandwich consists of layers of different pure ceramics, often stacked without order, but each chemically bonded to the next layer. These chemical bonds are not uniform in strength so the sandwich can come apart in much the same way a luncheon sandwich does. It is an example of a naturally occurring composite and even Mother Nature has her design problems.

Besides plates, one can imagine rings, strings and needles. Ceramics exhibit all of these shapes. The shape itself has been used

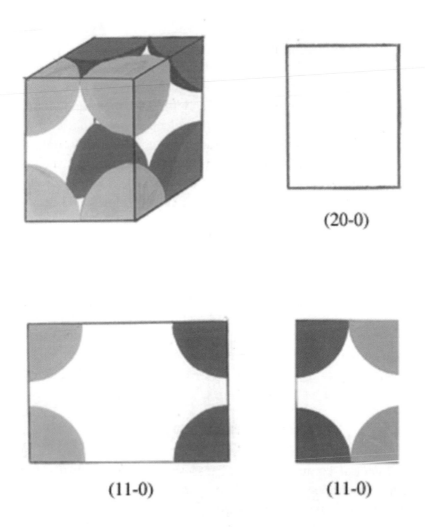

(20-0)

(11-0)   (11-0)

Figure 1-10 SCHEMATICS OF THE HEXAGONAL STRUCTURE

These schematics show the unit cell surrounded by three of its planes. In this case the unit cell is neither cubic nor right-angled. The (11-0) and (11-1) planes are given in two different versions due to the non-cubic nature. The angular relationships and relative lengths are shown in the diagram on the facing page. Note that the odd length side is marked with a small c rather than the expected small b which would be consistent with labeling the other sides with a small a. This convention is observed by X-ray diffractionists who determine most of the structure of ceramics. Diffractionists also use a system of four numbers rather than three to designate planes and directions for a hexagonal structure. This extra number, based on the addition of two of the usual numbers, has been designated here with a dash. The diagram, which is below the (11-1) plane diagram, is the concatenation of three unit cells. In this configuration it is clear why the structure is called hexagonal.

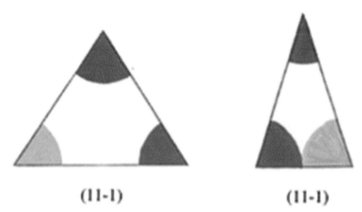

(11-1)   (11-1)

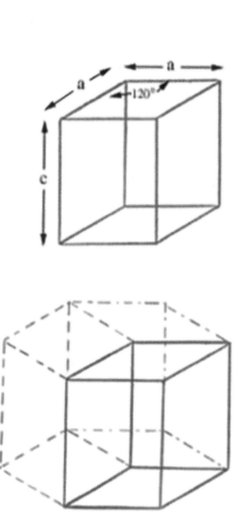

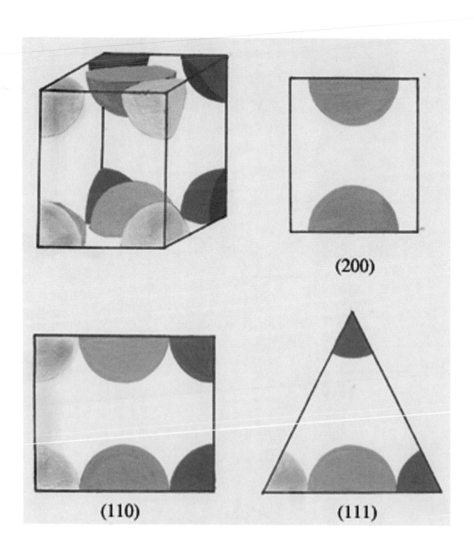

(200)

(110)   (111)

Figure 1-11 SCHEMATICS OF THE ORTHORHOMBIC STRUCTURE
The orthorhombic structure has right angles but the sides are not the same lengths.
This structure displays all four variations; simple, body centered, face centered and
base centered. A base centered orthorhombic structure is pictured where the top and
bottom planes contain a site in the center.

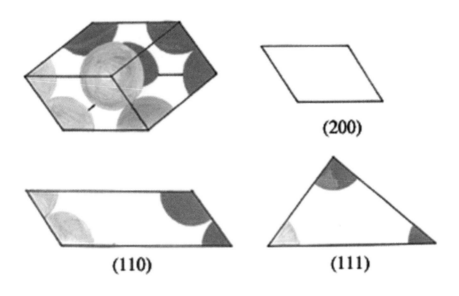

(200)

(110)   (111)

Figure 1-12 SCHEMATICS FOR THE MONOCLINIC STRUCTURE
The monoclinic structure has sides that are not the same length. In addition, at least
one corner is not a right angle. The (200) plane displays this variation in right angle
structure the most clearly. Monoclinic simple structure is shown here. A base centered structure for monoclinic is also possible. The least symmetric structure, triclinic, differs only from the monoclinic in that none of the angles are right angles.

to categorize ceramics; an example of this is asbestos. Asbestos forms a group of ceramics that are not the same chemically but all display a string-like structure.

MULTI-CRYSTAL STRUCTURES

Within a melt there are normally many nucleation sites, each of which will start a single crystal growing. Eventually these single crystals will impinge on each other. The result will be a miss-aligned interface as the probability of perfect alignment is very small between randomly generated individual crystals. Each individually generated crystal, or grain, is identifiable after growth has been completed. Where the crystals come together is the grain boundary. Figure 1-13 shows a schematic cross-section of grains and grain boundaries.

Crystal growth in controlled direction and location can be done if great care is used. Single crystals can be grown in this way and have many specialized uses. Multi-crystalline materials with alignment of the crystals in a specific non-random array are also grown. Most of the applications for these controlled crystalline structures are for properties that are sensitive to the packing factor in the crystal.

An example of a polycrystalline material with a controlled crystalline structure is pyrolytic graphite. There are several techniques for growing this material but the result is that all of the grains are aligned in the same direction. The growth surface or exposed surface is that of the closely packed hexagonal plane. This behaves like a hard outer shell. Here man has devised a means of producing a ceramic with the desired planes exposed which nature does by processing such as the weathering of clay.

Although the various shapes that a crystal can assume have their interesting and useful attributes, the vast majority of ceramic materials are multi-crystalline with non-aligned grains. These usually assume the spherical shape as nearly as possible while still filling volume entirely. The spherical shape minimizes the surface area, which is ultimately an energetically efficient structure for a solidifying ceramic. While the grains tend to be spherical, they do not tend to have the same volume unless further processing is done. During solidification grains are nucleated at different times and grow at different rates resulting in an array of grain sizes in the solid.

CRYSTAL IMPERFECTIONS

As a crystal forms defects become an inherent part of its struc-

Figure 1-13 SCHEMATIC OF SEVERAL CRYSTALS IMPINGING TO FORM
GRAIN BOUNDARIES
A simple cubic ceramic similar to cesium chloride is used to depict the grains. The
metal atoms of the ceramic are shown as light circles and the non-metals are shown
as dark circles. Of the four grains indicated here, the top left and bottom right are in
the (110) orientation; the top right is in the (100) orientation; the bottom left is in the
(200) orientation. Where they impinge are the grain boundaries. Note that the grain
boundaries are several atoms thick and have sufficient alignment to support some
bonding across them.

ture. An atom fails to fill a lattice site and produces a vacancy. An incorrect atom occupies a lattice site and produces a substitution. An atom slips into an empty area of the lattice and produces an interstitial. Distortions through the crystal form dislocations and twins. All of these to one level or another effect the properties of the crystal.

Vacancies are an unavoidable part of any crystal above absolute zero. As the temperature increases some of the absorbed energy goes into the formation of vacancies. The number increases continuously as the temperature increases throughout the solid state. However, increasing the pressure will decrease the number of vacancies at any given temperature. They also tend to coalesce into small voids especially with increasing pressure. Vacancies are most commonly formed when atoms are entirely missing due to their migration to the surface. Bonded atoms tend to form vacancies in stoichiometric groups to maintain overall electrical neutrality. If an impurity atom of a different valence substitutes then electrical neutrality may be maintained by eliminating a neighboring atom. An extreme case of this occurs when electrical neutrality is dependent on a regular array of vacancies because the coordination number of the smaller atoms will not allow enough proximity to the larger atoms within a crystal.

An increase in vacancies results in a decrease in density and may result in a decrease in strength. However, they are the primary defect within the crystal, which allows atom rearrangement on an inter-atomic basis. This supports diffusion and hole migration, which can prove valuable.

Dislocations are multi-dimensional defects, which can be seen in a transparent crystal. The screw dislocation is like a helix through the crystal. It is difficult to identify or draw. The edge dislocation is more easily visualized. It is basically an extra plane through the crystal, and its side can be represented schematically. Usually there is no need to differentiate between these two types of dislocations, as most dislocations are an intricate combination of both types. See Figure 1-14 for a schematic of an edge dislocation. This figure also shows interstitial atoms and vacancies. Dislocations are important in strengthening mechanisms and plastic flow of crystals.

Twins, like dislocations, are defects that extend through the grain. A slipping of the normal packing arrangement of the crystal such that a mirror image of the structure is created causes them. Figure 1-15 illustrates this. It would not seem that a mere slipping of

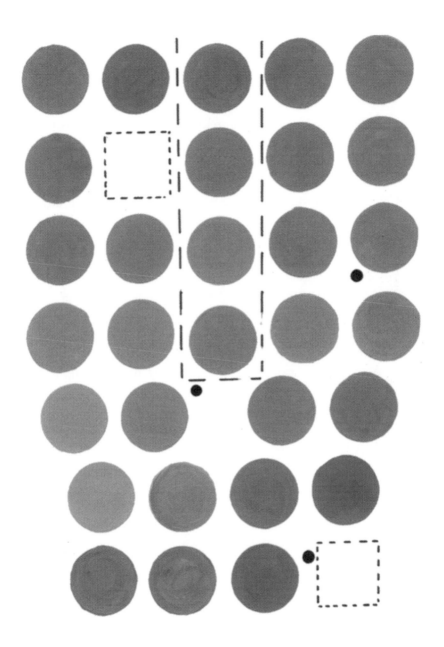

Figure 1-14 SCHEMATIC OF AN EDGE DISLOCATION
The structure shown is that of cesium chloride with the metal atoms in the (100) plane. The gray circles depict these. The small black circles indicate the interstitials, and vacancy sites are indicated by dotted lines forming a square. The edge dislocation is marked with a dashed line to show its outline, which is one atom thick; a thickness of one to two atoms is the most common.

the structure would cause much disturbance to the properties of the grain but, twins play a leading role in both fracture and plastic flow in many ceramics.

The grain boundary is usually the area of the greatest miss-alignment within the ceramic and consequently the area with the least chemical bonding density. As a result it is along grain boundaries that fracture is most likely to occur. Also, large impurities will tend to accumulate here. Grains are unavoidable within a ceramic of any size because there are limits to the growth potential of a single crystal. A grain boundary in its entirety can be seen in polished translucent gems of a few grains in size. The grain boundaries appear as cloudy areas, which disappear if viewed on their edge.

GLASSY STATE

When a ceramic is being cooled from the liquid state it is possible for the cooling rate to be sufficiently quick that nuclei either do not form, or do not grow. If this happens the liquid will produce an intermediate state of order, the glassy state.

The structure of the glassy state is closer to the solid state than the liquid state. Some short-range order of the crystalline type is apparent although a unit cell is not assigned to it. Figure 1-16 shows the unit cell structure of silicon dioxide in the crystalline state. From this drawing it is clear that even in the crystalline state silicon dioxide has a lot of open area in which rearrangement or distortion is possible.

The glassy state of a pure material has many of the mechanical attributes of the solid state. It can maintain a sheer stress up to a softening point, although it does not support sufficient diffusion to crystallize. The density is close to that of the solid state and so is the gas permeability. The glassy and solid states are especially similar optically.

LIQUID STATE

As a ceramic is heated from either the solid or the glassy state it normally reaches a temperature at which it melts and enters the true liquid state. The higher the pressure the ceramic is under during heating the more likely it will become liquid rather than passing directly into the gaseous state. All ceramics eventually become gases when the temperature is high enough and the pressure is low enough.

The liquid state shows the most diversity in structure of any of the states. On melting the structure is very similar to that of the solid

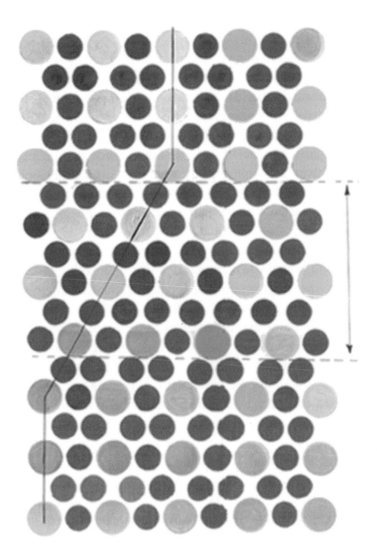

Figure 1-15 TWINS WITHIN CESIUM CHLORIDE

As in Figure 1-5, light circles indicate the cesium, the chloride is indicated by the dark circles. Because this is a (112) plane the cesium is cut through the center and appears as a single circle for each unit cell involved. The chloride is segmented in this plane such that three smaller circles correspond to one chloride atom in a unit cell. The dashed lines show twin boundaries, and the arrow marks the twinned region. Although it is possible for a twin to exist singularly and not transverse the entire grain this is unusual. Most grains that display twinning have the twins transverse the whole grain and exist in at least pairs. Twins come about primarily from mechanical working rather than as a growth fault on crystallization. They are seen only in specific planes and structures, which lend themselves to planer slip while maintaining bonding.

or the glass from which it came. As the liquid approaches the gaseous state there are a series of small structural changes often accompanied by second order transformations.

Melting relieves any distortions incurred by stressing the ceramic in the solid or glassy state. This plus the relaxing of the structure account for most of the volume increase due to melting. This increase is usually in the range of 2%.

Although the atoms involved may move very little the bonding changes dramatically on melting. The major change is the breaking of chemical bonds which held the structure in a three dimensional array before melting. It is this bond breakage that allows portions of the ceramic to slip easily over other portions of the ceramic and produce the mechanical behavior of the liquid.

On melting the atoms regroup into electrically neutral clusters which usually maintain stoichiometry. These clusters can vary widely in size and composition from one material to another. However, for a specific material they are quite uniform. As the temperature increases the size of these clusters continues to decrease. Eventually there will be some free atoms, which will be reflected in a measurable vapor pressure over the liquid. The measurable vapor pressure starts at a discrete temperature and increases with increasing temperature. This corresponds to a second order transformation. An increase in pressure will have the same affect on the structure as a decrease in temperature. Similarly, a decrease in pressure will be comparable to an increase in temperature.

TRANSFORMATIONS ON HEATING

Starting with a perfect crystalline ceramic at absolute zero, in which there is no thermal or internal energy, one may mentally consider the affects of heat to this aesthetic but unattainable material. Initial heating allows the atoms of the crystal to vibrate around their equilibrium positions permitting distortion and rearrangement. Vacancies will be the first defect to appear. These will continue to increase with increasing additions of thermal energy. This trend in vacancy increase is never curtailed throughout the entire temperature range of the ceramic. However, changes in state make the existence of the vacancies less meaningful.

If the heating process is done very slowly, so that large-scale distortion of the ceramic is avoided due to a sharp thermal gradient, than the crystal avoids the affects of internal stresses. Internal stress-

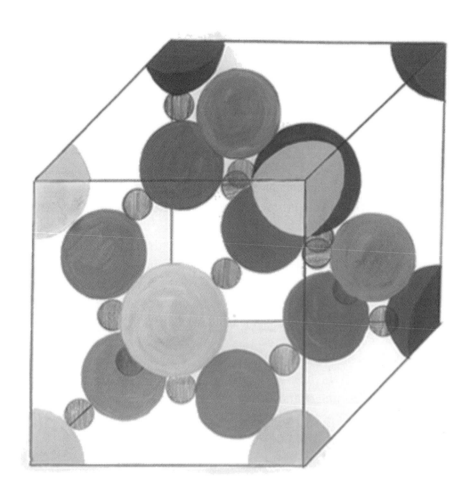

Figure 1-16 THE UNIT CELL OF SILICON DIOXIDE

This diagram shows the unit cell of silicon dioxide in the beta-cristobalite form, which is the stable phase of silicon dioxide above 405 C^O. It is face centered cubic with one silicon and two oxygen atoms associated with each lattice site. The large gray circles are the oxygen atoms. The small shaded circles are the silicon atoms with the vertically shaded circles being more forward than the horizontally shaded ones. The silicon-oxygen bonding is of an approximately 50% ionic and 50% covalent character with uniform bonding strength throughout the three dimensional structure. Although the bonds are not shown here, they can be inferred.

33

es are first manifested in the form of dislocations. Twins can also appear but grain boundaries will not be present because we started with a perfect crystal, which infers a single crystal. Only at high enough temperatures to allow significant diffusion is it possible for grain boundaries to form.

If the stresses associated with the heating have been high due to a rapid rise in temperature the result is a high defect structure with many dislocations. With sufficient mobility the crystal will reorganize into a multi-crystalline form in an effort to coalesce the defects into the grain boundaries. Because of the restraining affect the necessity for chemical bonding exerts, not all ceramics recrystallize to relieve stress. An inability to recrystallize is usually associated with ceramics that have angularity to their bonds such as carbides.

The perfect ceramic at absolute zero would not only have been defect free it would also have been in the most densely packed crystalline form possible for its composition. Heating decreases the density of the ceramic and the ever increasing number of vacancies allow for sufficient space that eventually the closely packed crystalline form of the lower temperature is not the best structure for the higher temperature. When this happens a phase change occurs and the crystalline structure of the ceramic converts to another type of crystal. Often this can be accomplished by bending or shifting a bond rather than a radical rearrangement in atomic bonding. Some phase transformation requires so little change that it can be accomplished almost instantaneously. There is energy associated with the transformation sufficient for it is considered a first order effect.

If the ceramic is at a point where it is about to change crystalline phase the presence of internal stress will effect the temperature at which this transformation occurs. Internal stress can both lower and raise the transformation temperature. In most cases the presence of stress will lower the transformation temperature because the rearrangement of the atoms during transformation tends to remove distortion. If the defects are such that they block the shifting of the atoms necessary for rearrangement into the new crystalline form than the phase change temperature will be increased. It is possible to have these two tendencies balance each other, which results in no measurable change in transformation temperature.

Most simple ceramics display three phase transformations in the solid state. With increasing temperature, each successive phase is less dense than the previous one and is more closely aligned with the liquid it is to become. Although melting has very noticeable affects on the mechanical properties of the ceramic it is not any more dramatic in atomic rearrangement than a phase transformation. Within the liquid state there are rearrangements in structure as there are phase changes within the solid state. The major difference is the degree of energy associated with these changes. Liquid changes are considered second order transformations. When a transformation renders the liquid insoluble with its previous structure it is given a phase like designation.

As in the transformation of the ceramic from a solid to a liquid, the transformation of a ceramic from the liquid to a gas is much more noticeable from a mechanical point of view than a structural one. The liquid is already discretely molecular in structure prior to vaporization. Transformation is often difficult to pinpoint except for the accompanying energy absorption. Most ceramics form a dense gas at the gas transformation temperature, which eventually behaves like an ideal gas as the temperature increases. The transformations that occur in the gaseous state include the breaking down of the molecules into atoms. These transformations are similar to the liquid state in that they have only second order energy levels associated with them.

TRANSFORMATIONS ON COOLING

If cooling were conducted sufficiently slowly the transformations which occurred on heating would be reversed. That would not only try the patience of most processing engineers, it would also avoid some interesting and useful structures. Condensing gas is not a good way to reconstruct a ceramic except of a simple structure. Only graphite is commercially produced from the gaseous state and then not directly from carbon atoms.

Transformation from the liquid state is of significant interest. The structure right above the melting point is sufficiently similar to that of the solid state directly below the melting point, there is no question as to the resulting structure on solidification. However, solidification can be avoided by rapid cooling to form a glass. The temperature range in which a glass is formed is usually listed as the median temperature in the range, the glass transition temperature. The glass transition temperature is often interchanged with the fictive temperature because the values correspond although the definitions differ. Figure 1-17 shows the volumetric relationship between glass transition and solidification.

Once a glass has formed it will remain a glass on cooling because the rearrangements necessary to allow solidification become more and more difficult as the structure becomes ever more ridged. Decreasing temperature makes glass increasingly deviant from the equilibrium structure. For this reason many glasses have a lower temperature limit below which they are not viable materials. At this minimum temperature the glass becomes so distorted trying to assume its equilibrium crystalline form that it is extremely brittle. This is sometimes referred to as the brittle transition temperature although it is really a broad band of temperatures. Eventually, given enough time and a low enough temperature, the ceramic will return to its crystalline form although it may have shattered into shards to do so.

DIFFUSION

In order to create changes in structure and composition, the atoms within a ceramic need mobility. One mechanism of mobility is diffusion wherein individual atoms move from their original sites. This movement does not necessarily result in a net flow of one type of atoms in a specific direction. Because most atoms in a ceramic are chemically bonded, the bonds must be broken before movement can occur. The broken bonds are usually reformed at the new location. Electrical neutrality must also be maintained except in a localized area or an applied electrical field.

Generally diffusion makes use of inherent defects within the material to allow motion. Atoms move much more easily over the surface of a ceramic than through its bulk. Similarly, atoms move more easily along grain boundaries than through the grain. Less densely packed planes promote diffusion over more densely packed planes within a grain. Even within a loosely packed plane, diffusion is very slow without flaws like vacancies and dislocations. In addition Frenkel defects, Schottky defects and doping augment diffusion.

A non-stoichiometric composition of the ceramic also aids diffusion. The extra atoms associated with the deviation from stoichiometry sprinkled through the ceramic are only loosely bonded. What bonds they have must be at the expense of bonds otherwise given to the even, regular bonding of the stoichiometric ceramic. Between the weakening of the overall bonding due to the extra atoms and the physical distortion of the crystal due to the extra atoms, the ceramic is more capable of rearrangement.

In a stoichiometric ceramic, atoms tend to diffuse as an elec-

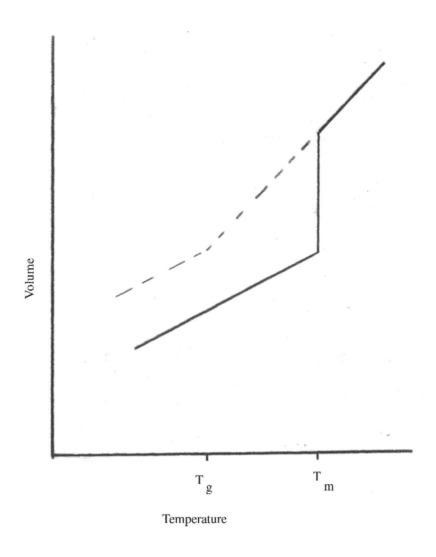

Figure 1-17 GRAPH OF GLASS TRANSITION VERSES SOLIDIFICATION
The solid line shows the volumetric changes associated with solidification.
Following the line from the upper right it can be seen that the cooling liquid has a
discontinuous volumetric change at the melting temperature, T_m. If the liquid is
cooled sufficiently rapidly that the ceramic does not solidify, the material becomes a
super-cooled liquid. This is depicted as a dashed line following the same slope as
the liquid line. The super-cooled liquid becomes a glass at the glass transition temperature, T_g.

37

trically neutral cluster. The size of the cluster makes diffusion very slow. A comparison of a stoichiometric and a non-stoichiometric ceramic can be seen in Figures 1-18 and 1-19.

The driving force behind net diffusion in a specific direction is chemical activity. The activity can be approximated by composition variations. For example, if a ceramic has a higher concentration of oxygen at one end of the material than the other, then diffusion will try to produce a net flow of oxygen to the side with the lower concentration. Generally the flow is not appreciable at temperatures below one half of the melting point of the ceramic. However, even at low temperature the flow will be maintained especially with large concentration differences and highly flawed structures. A large concentration difference for an oxygen bearing ceramic occurs when the ceramic in a vacuum. The ceramic may be without an internal oxygen concentration difference but it has a very big difference between the surface, which has no oxygen, and the bulk which does.

Diffusivity, the rate at which diffusion occurs, increases as a function of increasing temperature. A graph of temperature verses the log of diffusivity result in straight line for a specific structure. The slope of the line becomes greater as the structures become less dense. Sometimes there is also a discontinuity in the plot such as at the glass transition temperature. These discontinuities usually occur at non-equilibrium structural changes. An important application of diffusion is diffusion bonding; a serious consequence of diffusion is corrosion.

IRRADIATION EFFECTS

When irradiation energy strikes a material individual atoms become excited into higher energy states which render them more chemically active. If the material is not in an environment that will allow reaction of these excited atoms than the energy will slowly dissipate and the material will return to its pre-irradiated state. Irradiation energy can also move atoms out of their original position. This happens mainly when the dose is high and the atoms are light in atomic weight.

Irradiation damage due to energy bombardment is obvious in transparent glasses because it induces color centers. Ceramics can also have color centers induced. However, the more likely effect of energy bombardment of a ceramic is the chemical reaction of the affected atoms. Irradiated graphite produces cross-linkages reminiscent of a polymer.

All of the effects produced by irradiation energy can also be produced by irradiation particles. Because of the mass involved in irradiation particles, the atoms struck are much more likely to be moved out of position and split into daughter products. Within a ceramic, displaced atoms are usually found in nearby interstitial sites. Within a glass, displaced atoms are moved further from their original site due to weaker bonding. The displaced atoms of the glass tend to remain in the vicinity of their original site to maintain electrical neutrality.

The more tightly bound the atom is, the less likely it will be knocked out of position by irradiation of either type. Atoms in a glass are more likely to be moved than atoms in a ceramic. Atoms at grain boundaries are more likely to be moved than atoms within the grain. Similarly, atoms within the grain are more likely to be moved if they are near a defect like a void than if they are more perfectly bound. A common result of high dosage and/or sustained irradiation is that grain boundaries can be sufficiently damaged to produce grain pull out.

Although the atoms may be knocked out of their position by irradiation they may return to their original site if they have sufficient mobility. At low temperature and under high bombardment this does not happen both because of slow mobility and the rapid build up of displaced atoms. The result is clusters of vacant lattice sites. At high temperatures the atoms have sufficient mobility via diffusion to return to their original sites and can keep up with the irradiation bombardment displacement up to modest doses.

The formation of daughter products greatly complicates the situation for the recovery of the material to its pre-irradiated condition. The daughter products are less dense and more numerous than the original atom. Thus, several larger, lighter atoms at the same location replace a small dense atom. This major increase in volume produces tremendous space problems within the material. The production of gases as a by-product further stresses the material unless the material is permeable to that gas. Fortunately most irradiation damage is on the surface and daughter products are free to collect there without causing catastrophic stress build up.

GAS PERMEABILITY

In general ceramics have strong resistance to gas flow. Raising temperature and gas pressure increases gas permeability. Adding pathways such as grain boundaries also aids permeability. Usually

Figure 1-18 DIFFUSION IN A STOICHIOMETRIC CERAMIC

The ceramic depicted on the facing page is comprised of a metal, shown as the light circles, and a non-metal, shown as the dark circles. This is a slice taken through a close packed plane where the only defects are seven vacancies. The arrows and letters are to identify possible paths for the atoms. Location A is a vacancy of a metal atom and is therefore too small for a non-metal to move into easily. Similarly, location B is a vacancy of a metal atom. Both A and B could accommodate any of the neighboring metal atoms. From the diagram it may appear impossible for the metal atoms to squeeze between the non-metal atoms to reach the vacancy. In fact, the atoms are not required to move directly into it via the plane shown. It should also be remembered that although atoms are shown as solid balls and usually behave as a solid mass they are mainly empty space, which only requires a shift in the orbital of the electrons to "squeeze" though to another site. Locations C and C' are equally accessible to the metals atom marked. Since the atom cannot respond to the additional vacancy of C", it is just as likely to move into C as C'. Given the probability of C" being empty once the metal atom has moved, its existence is only of visual significance. Location D is a pair of vacancies, which would make the movement of a non-metal fairly easy. Vacancies often appear in small clusters both to minimize energy levels and to maintain localized electrical neutrality. In this case the movement of the non-metal is in the opposite direction to the other arrows. With no concentration gradient, there is not a net flow of atoms of either type. Location E is a non-metal vacancy, which is large enough to accommodate either type of atom. Although the metal atoms in the vicinity of the vacancy may relax into the space slightly, they will generally not occupy it due to their inability to form a chemical bond with a non-metal. Most likely a non-metal will occupy this space and bond with the surrounding metal atoms (see facing page).

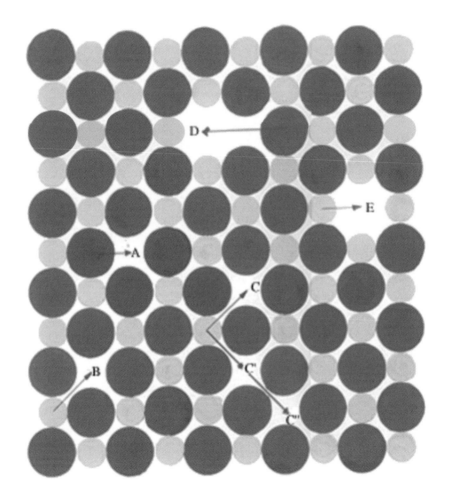

crystalline ceramics are less permeable than glasses of the same composition because of their greater density. However, glasses can sometimes display equally low gas permeability due to a lack of grain boundaries.

The size of the gas atom or molecule effects the permeability. Helium can slip through most ceramics whereas oxygen needs a path. Oxygen is larger than helium and becomes larger still if it forms a pair. Additives augment the natural pathways of polycrystalline ceramics. They tend to collect in the grain boundaries and further widen the space between grains. Active gases like oxygen can preferentially react with the additives or the less tightly bound atoms in the grain boundaries. Reactions of this type produce channels through the ceramic, which render it highly permeable.

CORROSION

As a group, ceramics are the most resistant materials to chemical degradation, or corrosion. Their chemical bonding makes any further chemical reaction less likely as the original bonds must be broken before new ones can form with the corroding substance. This means that the corroding substance must be more chemically active than the atoms within the ceramic. It also means that the corroding substance must be in intimate contact with the ceramic at the time of the bond breakage or the ceramic bonds will reform.

The amount of corrosion a ceramic sustains in a particular environment is very dependent on the structure of that ceramic. Ceramics that have an open structure and allow the penetration of the corroding agent suffer greatly whereas the same material with a closed structure may be unaffected. As in gas permeability, grain boundaries and channels lead to areas of high corrosion. In addition, internal stresses or defects decrease the strength of the chemical bonding within the ceramic and increase the likelihood of corrosion.

The nature of the corrosion itself can limit the corrosion rate. The products of corrosion are voluminous and, if trapped within the ceramic, form a barrier to additional corrosion. Once these corrosion products are in place any additional corrosion will require the corroding agent to diffuse through the product. Corrosion product on the surface can also limit corrosion rate if the product is highly adherent.

If the corrosion products are retained within the ceramic, the ceramic may stay the same size or expand during corrosion. Thus a corroded ceramic will not be as readily identifiable as other similarly

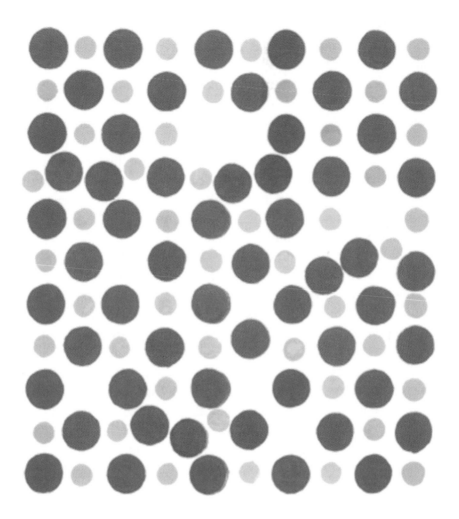

Figure 1-19 DIFFUSION IN A NON-STOICHIOMETRIC CERAMIC
This figure is the same as Figure 1-18 with the exceptions that it is of a less closely packed plane and there are four additional non-metal atoms. Even in a loosely packed plane, the addition of the larger non-metals causes more spatial problems in the crystal than the addition of the same number of smaller metal atoms. It is for this reason that the amount of deviation from stoichiometry in a ceramic which maintains a crystal structure of an ordered array is less in the direction of the non-metal composition variation than in the metal variation. The vacancies have not been marked so that there is no prejudicial arrow to consider atomic movement. This coupled with the added space of this crystallographic plane and the distortions due to stray extra atoms makes diffusion seem more eminent than it did in the closely packed plane of Figure 1-18. Rigidly ordered stoichiometric structures and closely packed planes both greatly reduce diffusion. The former is primarily due to extremely stable bonding conditions that do not support the distortions diffusion causes. The latter is more easily considered based on space alone.

damaged materials which use weight loss as an indicator. Testing to insure original properties is used to verify a lack of corrosion.

Impurity atoms and additives along the grain boundaries enhance the possibility of corrosion products being produced away from the surface of the ceramic. Processing to control the quantity and location of impurity and additive atoms can greatly enhance the corrosion resistance of a ceramic. If impurity and additive atoms cannot be removed from the ceramic then it is desirable to remove them from the grain boundaries. Within the bulk of the grain they are more difficult to reach by the corroding agent. Locating impurity and additive atoms within the bulk of the grain also improves the mechanical properties of the ceramic.

The absence of grain boundaries in glasses eliminates the concern over impurity and additive atoms being located there. However, glasses have sufficiently open structures that these atoms can still come into contact with the corroding agent. If this happens the corrosion can pervade the entire bulk of the glass. Fortunately most glass is transparent and this kind of corrosion will be obvious due to the change in optical properties that will result.

In anticipating the effects of corrosion it is important to consider it mainly a surface phenomenon with the most severe damage at the interface between the material and the corroding environment. If the corrosion product formed on the surface is volatile or non-adhering then there will be no barrier formed to on going corrosion at the initial rate.

It is also important to bear in mind that corrosion occurs under very specific conditions. Even if the corroding agent is in proximity to the ceramic it may not corrode. For instance, sulfuric acid will corrode cordierite ceramics when the sulfuric acid is in the liquid state but not when it has been heated to the vapor state. This is a clear exception to the generality that chemical reactions proceed more rapidly with increased temperature. Here a change of phase has altered the chemical reactiveness of the material in the reverse of that expected.

With polycrystalline ceramics a paradox exists. Ceramics with a few large grains rather than many small grains would appear the most desirable because they have fewer grain boundaries along which corrosion can proceed. In fact, the opposite obtains. Large grains mean that only limited paths exist through the ceramic and so corrosion is concentrated here. Small grains mean both more paths through the ceramic and less miss-match between the grains. Thus the corrosion rate along the large grain boundaries is larger than the corrosion rate along the grain boundaries of finer grained crystals for the same conditions.

If corrosion is completely around a small grain, it will fall out of the bulk but not too much material will have been lost. If this happens to a ceramic with only a few large grains than every grain is a major loss to its integrity. As a result, carefully produced fine grained polycrystalline materials show the best large-scale corrosion resistance of any ceramic of that composition except a single crystal. Given the rarity and cost of large single crystals like the Hope diamond, their corrosion resistance is not usually a technical consideration.

COMPOSITION MODIFIERS

ADDITIVES

When a material is less than ten percent of the total weight of the host ceramic it is usually considered an additive rather than a separate component. Often more than one additive is introduced at the same point of processing. They can simply be mixed into the ceramic with no chemical bonding between the ceramic and the additive. Becoming a chemically bonded part of the ceramic is also possible. Additives vary from large enough to replace a grain of the ceramic to small enough to fit within an interstitial site. Just how and where additives are located, as well as how and to what extent they are chemically bonded to the ceramic, influences the effect the additives will have on the overall properties.

IMPURITIES

Impurities can be considered uninvited and unwanted additives. Because they are generally at the atomic level they are often difficult to eliminate. Often the cost of processing can not justify the degradation to properties caused by them. One solution to the impurity problem is to use an additive that will mask or counteract the impurity. This is particularly prevalent in the glass industry where impurities impart unwanted colors. These impurities can be masked by additives that produce complimentary colors.

PRECIPITATES

Precipitates form during processing, chemical attack, changes in temperature, or changes in pressure. If the material needed to create the precipitate is already within the ceramic then changes in conditions will allow it to form as an identifiable second phase. This type of precipitate often forms within the grain. If some of the material needed to create the precipitate is added after the ceramic is formed then the needed material is usually introduced via diffusion. This type of precipitate will tend to form at the grain boundaries.

When chemical bonding occurs between the precipitate and a glass there is a fair amount of latitude for the orientation of each because strict rules of crystallographic orientation do not apply.

However, if the host ceramic is crystalline than the way in which the precipitate bonds will adjust such that the crystal can maintain its array and still meet the angular needs of the bonds to the precipitate. It is further complicated if the precipitate is also crystalline. In this case there will be only one stable orientation in which the precipitate can form relative to the crystallographic orientation of the host ceramic. As a consequence all crystalline precipitates within a crystalline ceramic grain will have the same orientation.

INCLUSIONS

Inclusion is a general word used to include material which is not a major component. There is no connotation of desirability. In most cases an inclusion has both negative and positive attributes. Small sized, undesirable inclusions are more precisely called impurities. If the material is desired and deliberately introduced then the term additive is better used. Similarly, the term precipitate may be used if the material forms or reforms during processing. The correct use of the terms additive and precipitate requires knowledge of the process by which the material was introduced. Thus, these terms are more often used by process engineers. The more global term of inclusion tends to be used by applications engineers.

FILLERS

Fillers are additions to a ceramic used to reduce cost. There is no specific amount or location associated with them. The consequences of filler addition are usually negative. However, in the area of weight reduction, they can produce an improvement. As an example, a common filler to concrete is plant matter. The organic material gives the concrete initial strength while the concrete is forming but adds no long term strength after it rots.

SINGLE COMPONENT SYSTEM

A single component system behaves in a homogeneous way and has properties specific to it. A ceramic made up of a single element is always a single component system despite many coexistent phases. A single component system may have additives of up to ten percent by weight and still behave as a single component. Stoichiometric ceramics are single component systems because a molecule of the stoichiometric composition occupies each site within the unit cell. If several separate ceramics, each of stoichiometric composition, interlace such that a regular array is formed then this will behave as a single component system. The unit cell will include all of the distinct types of these stoichiometric compositions. As an example, aluminum silicate is made up of alumina and silica in a regular array without being striated into layers or agglomerated into clumps. It behaves like a single new material with identifiably different properties from either alumina or silica.

MULTI-COMPONENT SYSTEMS

When two or more ceramics coexist in a non-homogeneous way, they form a multi-component system. The simplest example is a mixture of the ceramics such that there is no chemical bonding between the clearly identifiable single components. Multi-component systems can also form in a layered structure. In this arrangement each layer has its own composition and unit cell with internal chemical bonding. Each layer may be bonded to the next layer chemically but this bond is not necessarily a primary bond. Often layers are held together by secondary bonds. Multi-component ceramics also exist as grains of single component ceramics adjoining other grains of a different composition. The scale of this arrangement may be enlarged such that a cluster of grains of like composition may be next to a cluster of grains of a different composition. Just when multi-component system ceases to be a ceramic and becomes a composite is usually a function of the processing needed to produce it rather than the nature of the final result.

CONTINUOUS COMPONENT

In a multi-component material the relative amounts of each component is not usually as important as how the components are distributed. If a component forms the only continuous path through the material it will dominate many properties even if the continuous path is only a narrow ribbon. For instance, a component that forms along the grain boundaries of a host ceramic such that the grains are completely surrounded determines the corrosion resistance of the ceramic. Figure 2-1 shows a schematic of this structure. Similarly, if a second component is completely contained within the grains of a continuous component than the second component will have little effect on path sensitive properties such as diffusion. Figure 2-2 shows a schematic of this configuration. Layered structures can also be continuous component materials in specific directions. This is

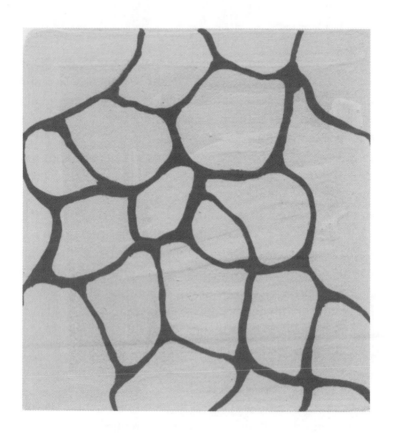

Figure 2-1 CONTINUOUS COMPONENT ALONG THE GRAIN BOUNDARY
This is a schematic cross-section of a two component system where the second
component has formed as a film around the individual grains of the primary component. The primary component is depicted in light gray and the secondary component is depicted in dark gray.

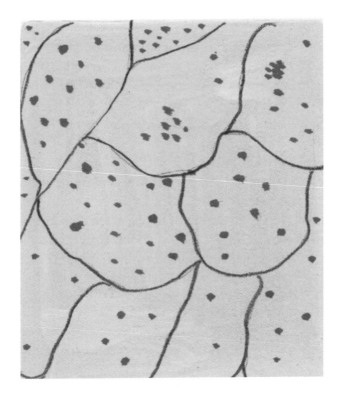

Figure 2-2 A TWO COMPONENT CERAMIC WHERE THE SECOND COM-
PONENT IS CONTAINED WITHIN THE FIRST COMPONENT
As in Figure 2-1, this is also a schematic cross-section of a two component ceramic. Here the primary ceramic is continuous and the secondary ceramic is completely surrounded by the primary. The light gray colored primary ceramic is
depicted as a crystalline material with grains. The dark gray colored secondary
ceramic is shown as a precipitate within the grains of the primary ceramic. The
thin lines are the boundaries between the grains.

illustrated in Figure 2-3.

The usual assumption for a continuous component material is that the continuous component is the only contribution to the property if that property is path sensitive. Many of these path sensitive properties are dependent on the cross-sectional area. If this is the case than only the cross-sectional area of the continuous component is used in the calculation. When more than one continuous path component is present in a ceramic, all of the continuous path components are considered. Each is given a proportion of the value calculated relative to the cross-sectional area of each path. If the cross-sectional areas are irregular and an average is difficult to obtain, than the volumes of each continuous path component are used as the proportional factor.

INTERSPERSED COMPONENTS

The most common arrangement for a multi-component system is interspersed components, illustrated in Figure 2-3 and Figure 2-4. The principal disadvantage of interspersed components is bonding weakness between the components. Strength can be augmented by selection of components that bond well or by additives that produce bridging bonds. Usually sintered ceramics should be considered an interspersed two component system with the second component having the properties of air.

CONCRETE, A MULTI-COMPONENT CERAMIC

Concrete is an example of a complex multi-component ceramic. The materials that comprise concrete are primarily in their natural form and do not have simple compositions or structures. Furthermore, concrete is often reinforced by steel bars.

By definition concrete contains stones. These can vary in size from fine sand up to large rocks. Usually the sand is beach sand of high silicon content; the rocks are more diverse in composition. Both are primarily crystalline with a small grain size. If they are crushed, the surfaces become rough which facilitates the mechanical bond with the interpenetrating cement. This augments the final tensile strength of the concrete. Most rocks and sand are not crushed because this process adds to the cost of the concrete and does not significantly add to its compressive strength. Compression is the primary loading mode of concrete used in construction.

Air is introduced during the mixing of the rocks, sand and

Figure 2-3 A TWO COMPONENT LAYERED CERAMIC

This schematic shows the edge of a two component layered ceramic. Although the layers are shown as alternating between two components the application could be extended to any number of different components in any order of layering. In the horizontal direction there is a continuous path through the entire material for each of the components.

cement. Some excess water is also introduced to insure that the cement has enough water to complete the chemical reactions necessary for it to fully set. Cement sets into a primarily amorphous ceramic with precipitates of crystals throughout the material. The fairly open structure allows for the easy escape of air and the evaporation of excess water. Cement can be considered a continuous component intertwined with narrow continuous pores.

Although cement does not chemically react with the rocks or sand in the concrete, it may react with the iron oxide on the surface of steel reinforcing bars. Iron oxide is chemically bonded to the steel from which it formed. Thus, iron oxide can sometimes serve as a bridge between the steel bar and the cement. However, the usual form of iron oxide that forms at low temperature, rust, is so defect ridden that it has negligible strength. A cross-sectional schematic of concrete is shown in Figure 2-5.

PHASE DIAGRAMS

Components are depicted as a function of temperature and/or pressure by a phase diagram. For a single component material such as an element, the diagram will show phases relative to variations in pressure and temperature. When a two component system is diagramed the phases are usually shown as a function of temperature and composition. These could also be diagramed as a function of pressure and composition if the data exists. Generally a family of temperature/composition diagrams are plotted with the pressure of each noted. When the pressure is not noted it is assumed to be atmospheric. For a three component system the composition variations are graphed in a triangular plot for a specific temperature and pressure.

These diagrams are of obvious use in processing and can also serve a function in materials selection/design considerations. For instance, if an oxide is to be used in air, the effect of the possible addition of oxygen to the material can be anticipated by the phase diagram. However, whether or not the oxygen in the air will enter the oxide and change the composition cannot be determined by the phase diagram alone. In addition, the rate at which this happens may be sufficiently slow to render it immaterial. There are no diagrams to determine if the oxygen will enter. Between size considerations discussed in chapter 1 and calculating diffusivity it can be determined if the oxygen is likely to permeate the oxide. If it does permeate, how quickly it will transform the phase of the oxide can be estimated by the use of a time-temperature-transformation diagram.

PHASE CHANGES

Phase changes can take place in response to a change in either temperature or pressure. The phase changes brought about in this way represent changes in the crystalline structure within the solid state or changes from one state to the next. This is a rearrangement of the atoms to form a new structure without a change in the composition of the material.

Phase changes can also occur when the composition of the material is sufficiently altered. As an example, consider titanium carbide surrounded by a carbon bearing gas which is diffusing into the body of the material. Initially the ceramic can accept carbon without change because some carbon sites are probably vacant. This will continue until the saturation point for carbon in that phase. Carbon will then begin to collect. This generally takes place at the grain boundaries where there is more room for it there than within the bulk of the grain. If enough carbon collects it will become an identifiable second phase. Both the composition and extent of the new phase can be determined by checking the phase diagram. The diagram does not indicate where the new phase will form. Standard microstructural techniques can be used to see it.

All possible ceramics from one end of the phase diagram to the other can be produced. Another element could be added to the material further increasing the number of phases and possible precipitation sites. In addition, processing can be devised that produces non-equilibrium but stable structures and compositions. It is this wide range of structure verses composition possibilities that make ceramics so diverse.

POLYMERIC COATING

Polymers find extensive use as coatings on ceramics. The most common use is as a seal from gas penetration. A polymeric coating is also a cheap, easy way of producing the desired finish on a ceramic. A large array of colors and sheens are obtainable with little skill or processing. Coatings can vary from a thin film of paint to a thick layer which also adds structural strength.

The principle disadvantage to the use of a polymer coating on a ceramic is the lack of primary chemical bonding between the two

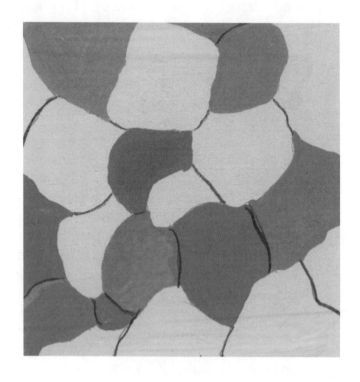

Figure 2-4 A SCHEMATIC OF INTERSPERSED COMPONENTS
This figure shows a cross-section of a two component ceramic where neither is
necessarily continuous. Within such a system the probability of one or the other of
the components having at least one continuous path through the bulk increases
with the volume percent of that component. If this figure were a photograph in the
actual size of a large grained ceramic, the path length through a phase could be
estimated via direct measurement.

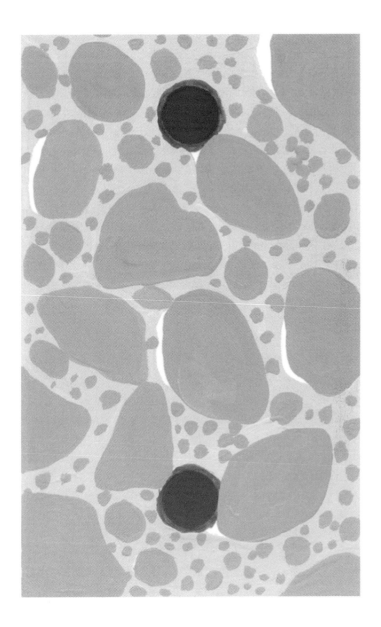

Figure 2-5 A CROSS-SECTIONAL SCHEMATIC OF CONCRETE
This figure shows a typical cross-section of reinforced concrete. The continuous light gray represents cement. The black circles depict the end of steel reinforcing bars with the usual rust around it shown as dark gray. Sand and stones are shown as medium gray; the white spaces are voids. Concrete is generally not gas tight so the voids contain air. The voids tend to coalesce around the stones because air has been trapped there during mixing and bonding is very weak in this location. The stones are strong in compression which contributes to the overall strength of the composite. Stones also serve to dissipate cracks that form in the cement.

materials. The polymer is completely internally bonded and will have no free bonds available to interact with the ceramic. Even if processing could be developed to alter this, the amount of overall interbonding between the polymer and the ceramic would be minimal due to the nature of the polymer. Major rebonding of the polymer to the ceramic would alter its characteristics and probably render it undesirable.

If the polymer is applied as a continuous sheet over the ceramic it will have its own internal bonding to keep it in place as well as the mechanical bond that usually results when a fluid polymer permeates a usually porous ceramic. There is always mass attraction between the polymer and the ceramic but this is not substantial due to the low density of polymers. Failure often occurs when the polymeric layer is scraped or pealed off. High temperature melts the polymer. Low temperature causes the polymer to crack due to thermal expansion which is dissimilar to the underlying ceramic.

METALLIC COATING

Ceramics can be covered with a metallic coating to enhance properties. Like polymers, the bonding is primarily of a mechanical nature where the metal deforms into the porous surface of the ceramic. However, unlike polymers, fairly extensive chemical bonding is possible between the ceramic and the metal if these are chosen to be compatible. An example of this is given in Figure 2-6. There is also the secondary bonding of mass attraction which could be stronger than the same type of polymer-ceramic bond due to the usually higher density of metal.

Metal coating provides a gas barrier for the ceramic but the color range is severely limited to the few colors in which metals occur. Metallic surface reflectivity will always be high which may be an advantage in some applications. Pure metals are subject to corrosion and may not last long in a harsh environment. The exception to this is a metal that forms an adherent surface oxide like aluminum.

In general a metal is more likely to be used as a reinforcing agent in a composite with a ceramic matrix than as a coating. Continuous wires or bars of metal could add considerably to the tensile strength of a ceramic while also increasing its thermal and electrical conductivity. When the metal is encased by the ceramic, the

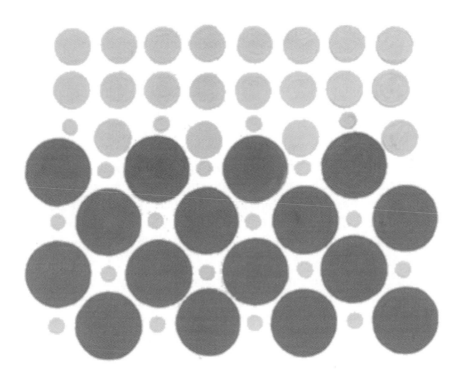

Figure 2-6 A SCHEMATIC OF A METALLIC COATING ON A CERAMIC
This schematic shows a metallic coating over a ceramic at the atomic level. The
ceramic is the face centered cubic phase of titanium carbide in the (100) plane.
Titanium atoms are shown as light gray circles and carbon atoms are shown as
dark gray circles. Above the titanium carbide is a coating of body centered cubic
titanium also in the (100) plane of its crystalline structure. Again the titanium
atoms are shown as light gray circles although larger because they are not chemically bonded. It is clear that in this type of matched arrangement, bonding
between the metallic titanium and the incompletely bonded carbon of the ceramic
surface could take place. In addition there will always be a metallic bond between
the titanium of the metal and the titanium of the ceramic. Similar coating of glass
is possible.

ceramic reduces or eliminates the corrosion susceptibility of the metal.

Metallic paint is not a true metal coating; it is composed of metallic powder suspended in a polymeric base. The paint has the color and reflectivity of the metal but the other properties of its polymeric base. This combination has proven to be an inexpensive and effective way of imparting a reflective gas barrier on a ceramic.

GLAZES

The principal use of a glaze over a ceramic is to enhance the surface appearance. Glazes also serve as a barrier to penetration of the underlying ceramic. The major concern is usually water but gases can also be blocked. Glazes may increase the strength of the final product. Like all ceramics, glazes are stronger and less susceptible to cracking under some compression than they are unstressed or under tensile stress. If they are in a state of compression due to processing, they add strength to the product both by the addition of their own strength and the inducement of a compression condition on the surface of the underlying ceramic. This compressive force retards the propagation of the small cracks inherent in many ceramics.

Most glazes are a glassy phase of silicon dioxide processed such that it is in compression. For the glaze to be adherent there must be some chemical bonding between the glaze and the ceramic. During thermal cycling the expansion of the glaze and the ceramic need to be well matched. Too much expansion of the glaze relative to the ceramic results in a reversal of the protective compressive finish to the glaze surface, which supports crazing. Too little thermal expansion of the glaze puts it into excessive compression and causes it to crack and peel. Additives are usually included to tailor the properties to suit the underling ceramic.

GLAZE ADDITIVES

Additives are included in a glaze to alter color more often than any other reason. Color modifying additions are generally ceramic and opaque. Most of the pigment added to a glaze is ceramic in nature because pure metals tend to interact with the glaze and change the color of both. Polymers generally decompose at the processing temperature needed for a glaze.

Ceramics take on a particular color due to their covalent bonding. The more covalent the bond is the more it moves from the ultraviolet range into the color range. Since the color of the ceramic pigment is dependent on its bonding, any disruption to that bond will alter or destroy the color. Pigments must not be so finely ground that they can be dissolved into the glaze and loose their characteristic bonding. Similarly, large chunks of pigment are undesirable as they diffract light which results in a loss of brilliance and a blotchy appearance.

A common additive to a glaze is a flux. Flux is a ceramic which acts to lower the melting temperature of the mixture and facilitate processing. Because the flux becomes a part of the finished glaze it must be considered along with any other additives as to desirability. Lowering the processing temperature will also lower the operating temperature.

CERAMIC COATING

Coating a ceramic with a ceramic is a type of simple composite. Usually the coating is chosen both to enhance the properties of the underlying ceramic and to bond well with that ceramic. Boron carbide on titanium carbide is an example. Both the coating and the underling ceramic contain carbon which serves as a bridge between them. The boron carbide adds strength to the titanium carbide while reducing the density of the whole.

Phase transformations due to the addition of an additive can cause a coating like arrangement within a ceramic. A mixture of zirconia and stabilized zirconia is made in this way. Similarly, ion exchange can take place at the surface to locally change the ceramic and cerate a coating. Sodium or Potassium ions will readily replace the aluminum in alumina; see Figure 2-7 for a schematic. Finally, chemical reactions can be devised that produce a wholly different ceramic on the surface where the reaction has taken place.

A chemical reaction approach to producing the coating results in an evenly bonded coating. However, it may not result in a coating that is in compression. The compressive state can be estimated by the relative volumes that the host and the final ceramics require. It is not essential that the coating be in compression but usually the strength is significantly improved if it is. Just how important the compression state is can be evaluated by considering the difference between the compressive and tensile strengths. In addition, the failure mode of the coating needs to be considered. For glazes both the strength and propagation of cracks is better in compression than in tension because glazes are based on silica. This cannot be considered more than a generality for ceramic coatings.

Multiple coating may also be useful but each additional ceramic adds to the matching problems. An example of successful multiple coatings can be found among the various combinations used on graphite whiskers. Graphite reacts with oxygen at elevated temperatures making a gas barrier coating desirable. Oxide coatings break down at high temperatures and react with the graphite if they are in direct contact. The usual solution is to use a carbide intermediate so the highly reactive graphite will be protected from both oxygen and the oxygen content of the oxide. A carbide coating alone would also be subject to oxygen reaction at a reduced rate. Further refinements on this are many, varied and proprietary.

TEMPERED GLASS

If glass is carefully cooled from its softening point such that it forms an outer coating in compression then it is considered to be tempered. Glass is more resistant to cracking, corrosion and thermal shock as well as stronger in compression. With tempering the problems inherent with other coatings of adherence or miss matching of thermal properties are eliminated. Tempering is used extensively because of its obvious advantages relative to its cost.

Glass is an interbonded whole. The compressive condition induced on the surface of the glass by tempering must be balanced by the interior of the glass being in tension. Crowding of the structure produces compression; spreading of the structure results in tension. Therefore, a direct correlation exists between the density of the glass and its stress level after tempering.

From Figure 2-8, it is clear that as the thickness of the compressive layer increases the corresponding tensile forces will also increase. After a point the increase in strength due to the compressive layer will be more than offset by the intensity of the tensile region. Each glass type and configuration has an optimum condition under which the tempering is the most effective. The general relationship between compressive layer thickness and increase in strength of the glass is given in Figure 2-9.

GLASS MODIFIERS

Glass forming ceramics are often hard to process due to their high melting point. They also have a limited range of glass formation before they revert to the crystalline state. Modifiers reduce the melting point due to their own lower melting point. In addition, modifiers will inhibit the glass former from crystallizing and increase the range of working temperature. The glass also becomes less viscous and more easily shaped.

Modifiers are usually oxides which interact chemically with the glass. The oxygen component of the modifier reacts to produce non-bridging segments in the structure. The metallic ion of the modifier remains in the open areas of the glass as a type of filler. Sodium oxide is a common modifier which reacts with silicon dioxide to form a distorted region which allows the sodium to remain loosely attached to the glass structure.

GLASS HARDENING

One means of strengthening a modified glass is to exchange the metallic ion of the modifier with a larger metallic ion. This can easily be done when the larger ion is more chemically active although any larger metallic ion will be effective. Usually the exchange is made at an elevated temperature to increase the rate of diffusion needed to make the exchange.

Initially the exchange is diffusion controlled as the modifier ion must move out of the way of the incoming replacement ion. However, as the replacement proceeds the stresses created in the swollen glass surface discourage further acceptance of the replacement ion. At higher temperature the glass is less dense due to thermal expansion and can accept more large ions. Usually only a thin coating of the larger ion is desired because it both induces internal stress and alters the optical properties. Once the desired exchange is complete the glass is cooled and the surface is then in compression due to the larger ions. This is depicted in Figure 2-10.

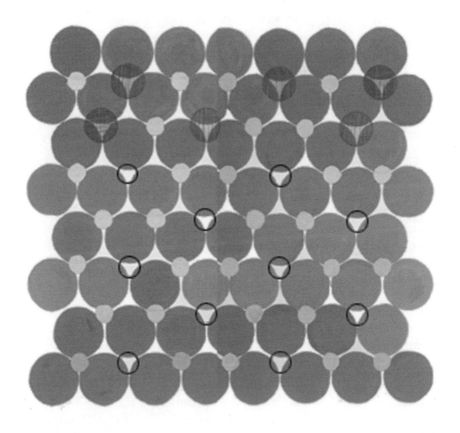

Figure 2-7 SUBSTITUTION TO FORM A COATING

This is a schematic of alumina in the [100] plane. The oxygen ions are represented by the dark gray circles in the basal plane. The light gray circles in a parallel plane above the oxygen ions are the aluminum ions. The dashed circles indicate were the aluminum ions are missing. Alumina has a regular array of aluminum vacancies to maintain electrical neutrality. Sodium ions are depicted near the surface. The sodium ions occupying an aluminum site are shown with a double cross hatching. The sodium ions occupying an aluminum vacancy site are shown with a single cross hatching. It takes three sodium ions to replace one aluminum ion and maintain electrical neutrality. Thus the sodium tends to enter the alumina in clusters of three although the alumina is stressed by this concentration due to the size difference between the aluminum and sodium ions. This stress causes a compressive condition in the region surrounding the substituting sodium ions. The chemical reactions used to produce this kind of substitution result in the sodium clustering near the surface and creating a coating. With sufficient time and/or high enough temperature the sodium will diffuse into the bulk of the alumina and the surface coating effect will be greatly reduced. Because diffusion is sluggish up to half the melting point this type of coating is useful at low operating temperatures. The process is also used to harden the surface of glass.

Compression   Tension

Figure 2-8 TEMPERED GLASS
The figure on the left depicts a cross section of tempered glass where the shade
corresponds to the density. Tempering is a process that produces a gradation of
density in the glass rather than a defined layer. The figure to the right shows the
stress condition of the cross section.

65

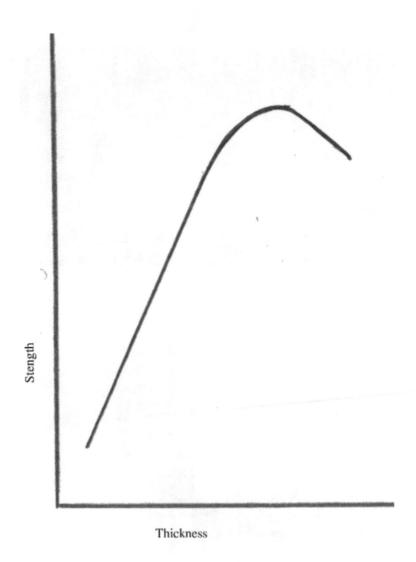

Figure 2-9 STRENGTH OF TEMPERED GLASS
The thickness of the compression layer of tempered glass is shown above relative
to the overall strength of the glass.

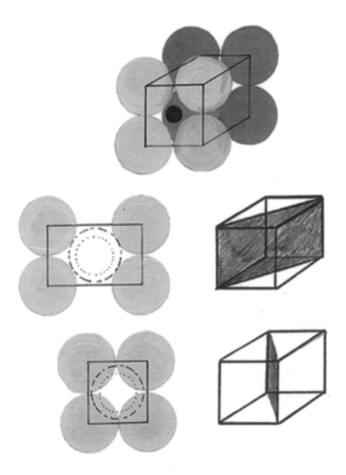

Figure 2-10 GLASS HARDENING BY ION EXCHANGE

While glass is not a crystalline substance, it only deviates from crystalline structure by a sufficiently small amount that it is more accurately envisioned as a
flawed crystal than as a random array. The top figure is the unit cell of hexagonal
silicon dioxide. The light gray circles are oxygen and the small dark gray circle is
silicon. Both the middle and bottom diagrams show the 110 plane of the unit cell.
From the middle diagram it can be seen that the center space can accommodate
either a sodium atom (shown as a dotted line) or a larger potassium atom (shown
as a dashed line). The bottom diagrams show the other direction of the 110 plane
cannot easily accommodate either of these atoms. Because the unit cell is opened
by the reaction of the silicon dioxide with the sodium oxide modifier, the sodium
is able to fill this center area which has been enlarged by the additional oxygen
residing on one side. This space will not be large enough for the potassium atom
without significant thermal expansion.

PHYSICAL PROPERTIES

DENSITY

Ceramics display a wide range of densities due to all the possible combinations of metal and non-metal compositions. Density is influenced by structure, phase and porosity. The values fall between the low density of polymers and the high density of very heavy metals. Density is at a maximum at absolute zero and continues to decrease with increasing temperature for an equilibrium structure. The decrease in density is discontinuous at changes in phase and state. This behavior is depicted in Figure 3-1. If the structure is a deviant from equilibrium or has high porosity, then density may decrease with increasing temperature.

Porosity is a major contributor to density variation within a given ceramic. Generally it is greater than the contributions made by phase changes and misaligned areas like grain boundaries. Processing is the most common source of porosity.

For many high melting ceramics the only means to form them is through powder compaction and hot pressing. If porous ceramics are exposed to high enough temperatures diffusion is effective at reducing porosity by grain growth. This will be augmented if a stress is applied. Grain growth with porosity decrease will continue until a non-porous, large grain crystal is formed. The practical limit is significantly short of this. Grains of larger than 20% of the bulk and porosity below 1% would be unusual. Generally for high strength applications such as cutting tools, porosity below 2% is achieved. Grain growth advances disproportionally quickly as compared to porosity reduction. Reduction in porosity results in major grain growth and significant strength reduction.

VISCOSITY

Relevant to glasses, viscosity varies with composition, temperature and internal stress in a non-linear way. Usually either an elongation test (ASTM C336) or a beam bending test (ASTM C598) are used to determine viscosity. Once viscosity data has been generated for a specific composition then temperature is used to track the viscosity. For example, a value like softening temperature is listed to indicate a specific viscosity.

Viscosity takes time to reach its long term value. On heating glasses the viscosity value is larger than the stable value for that temperature. Similarly, on cooling it is lower than the stable value. Reaching a stable viscosity on cooling can take as much as an hour and heating is slower yet. A hundred hours of "soaking time" is not unknown to reach a truly stable viscosity.

The values of viscosity have a huge range. This is exemplified by the following list for silica glass:

TEMPERATURE	VISCOSITY	CONDITION
956° C	$10^{13.6}$ Pa.s	Strain
1084° C	10^{12} Pa.s	Annealing
1580° C	10^7 Pa.s	Softening
1723° C	10 Pa.s	Melting

THERMAL EXPANSION

At or near absolute zero, a ceramic will try to assume the densest crystalline structure it can, given the constraints of chemical bonding. The ideal structure would be defect free and the volume unfilled by the atoms would be a minimum. This open area, or "free volume", increases with increasing temperature in a fairly linear manner for an equilibrium structure. Most of the increase in space that the ceramic experiences on heating is due to a relaxing of the lattice. Additional space will be created in the form of vacancies. With this expansion the equilibrium crystalline structure becomes more and more flawed and enlarged until a more ordered structure of lower density is favored. When this occurs the crystal will change to a less dense phase and a noticeable shift will occur in the thermal expansion along with a change in the slope. An even more noticeable shift will occur between states. The largest shift is the state change from the liquid to the gaseous. This behavior is shown in Figure 3-2.

Non-equilibrium structures such as the glassy phase will have thermal expansion values similar to the crystalline phase from which they are a variation. Structural defects such as grain boundaries and dislocations will have little effect on thermal expansion. Even small variations in composition, which do not alter the structure, will not alter the thermal expansion to a noticeable extent.

It would seem that thermal expansion is a relatively smooth function of temperature for a material over the whole temperature range except for phase changes. This will hold true for equilibrium structures or near equilibrium structures but it is possible to create structures by processing that are very different from the equilibrium structure. When this occurs, the thermal expansion will also be very different. Ceramics can display the largest deviation from equilibrium values of thermal expansion of any material.

This wide deviation in thermal expansion values is best exemplified by silica because it has so many equilibrium and non-equilibrium structures possible in the solid state. Glass-ceramics of multiple phase compositions have been developed that cover a large range of values. Some structures of silica have been devised that show a negative thermal expansion value over limited temperature ranges. Silica cannot show a negative thermal expansion value for its equilibrium structure. However, it can contract from a low density non-equilibrium structure and in so doing reduce the overall volume of the material. When this happens within a composite the recorded value of thermal expansion can become negative.

SPECIFIC HEAT

The value of specific heat approaches zero as the temperature approaches absolute zero. From this beginning, a plot of specific heat will be an ever increasing function of temperature until the liquid state for a stabilized glass. A crystalline ceramic will have this trend in the solid state with discontinuities at phase changes. The liquid and gaseous states will display the same trend with the accompanying discontinuities at state changes. As with thermal expansion, specific heat is not sensitive to variations in defect structure or small changes in composition. A plot of specific heat versus temperature is shown in Figure 3-3.

Although there is always a positive slope to the specific heat verses temperature curve within a structure, the value of the slope can change dramatically. At very low temperature the slope is small. As the temperature rises so does the value of the slope. Most of the high slope values for a ceramic will occur between low temperature and ambient. Above ambient the specific heat slope will be smaller but

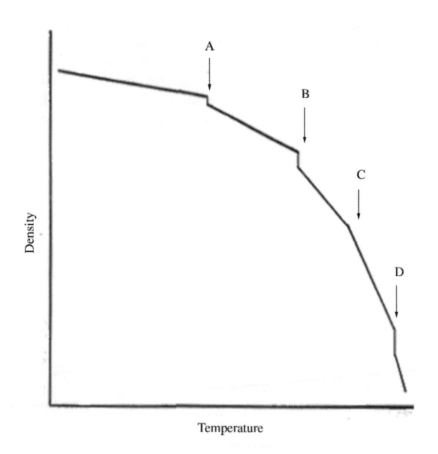

Figure 3-1 DENSITY AS A FUNCTION OF TEMPERATURE
This plot shows decreasing density for increasing temperature which is characteristic of an equilibrium structure ceramic. At very low temperatures ceramics display
highly dense crystalline structures such as face centered cubic. As they increase in
temperature they transform to ever less dense phases. Point A is at a phase
change. Point B is at the transition from the crystalline to the liquid state. Point C
is at a second order transformation within the liquid state. Point D is at the state
change from the liquid to the gaseous. Within the gaseous state the concept of
density becomes meaningless as the ceramic approaches the ideal gas condition.

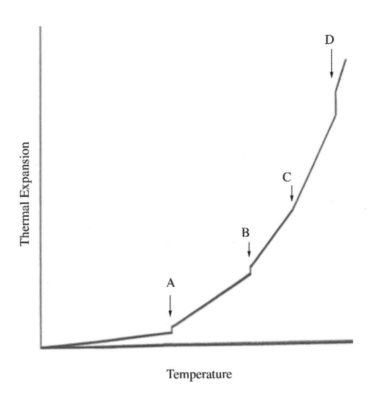

Figure 3-2 THERMAL EXPANSION AS A FUNCTION OF TEMPERATURE
This figure is closely related to Figure 3-1 and the letters are used at the same
positions of structural change. Both are basically a volume function. In this case
the values are based on a percentage change rather than an absolute number so the
values change less with increasing temperature than they do in Figure 3-1.
Otherwise, this plot would be simply an inversion of the density plot.

73

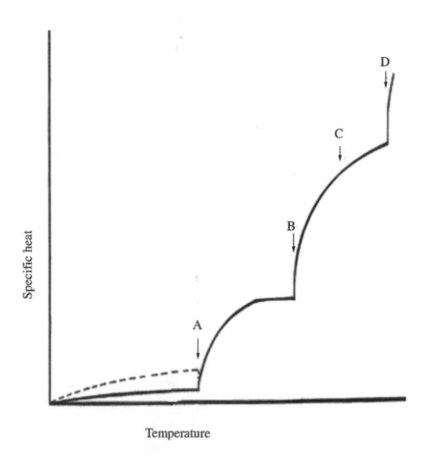

Figure 3-3 SPECIFIC HEAT AS A FUNCTION OF TEMPERATURE
The same general ceramic is assumed in this figure as was used in Figures 3-1 and
3-2. Similarly, the letters correspond to the structural changes. Specific heat tends
to show a smooth curve with a plateau for an equilibrium crystalline structure.
The region of the plot between points A and B show this behavior clearly. For the
liquid state the curve is an ever increasing one without any noticeable change at
second order transformations like point C. Glasses display a curve similar to the
curve segment between points B and C. A notable exception to this trend is alumina. It follows a trend that looks like the dotted section between the origin and
point A before it conforms to the general curve above point A. The reason for this
dip in the curve where no phase change is expected is probably due to the suppres-
sion of a low temperature phase, which would represent true equilibrium. At such
low temperatures true equilibrium may not be practical given the extremely sluggish movements of atoms.

usually not as small as at the very low temperatures. The slope of the specific heat will tend to remain at this small value throughout both structural and phase changes.

The amount of the discontinuous change that occurs with structure or state changes is not predictable. Specific heat may decrease with a structural change or melting rather than increase as would be expected. The extent of the change, either positive or negative is small for changes in structure or melting. However, when a liquid transforms into the gaseous state there is a large decrease in specific heat. This is on the order of half the value for the gaseous state as compared to the liquid state.

THERMAL CONDUCTIVITY

Thermal conductivity in a ceramic is accomplished primarily by transmission of energy through lattice vibrations. At absolute zero there is no lattice vibration and the thermal conductivity becomes zero. As the ceramic is heated lattice vibrations allow sufficient movement to transmit the energy necessary for some thermal conductivity. The ease of transmission of energy increases with increasing vibrations of the atoms until a point is reached where the increase in lattice vibration is balanced by the decrease in lattice order due to the accompanying thermal expansion. At this temperature the thermal conductivity will peak. Beyond the peak any increase in vibrational mobility of the lattice is more than compensated by the decrease in perfection of the crystalline structure. Other defects that the crystal might have will also reduce the thermal conductivity. Similarly, substitutional atoms, especially metallic atoms, will act as voids for the purposes of heat conduction. Eventually these factors will balance each other for the stable structure and the thermal conductivity will become almost a constant value. This occurs at a high enough temperature that phase changes will have little effect on vibrational mobility.

When the ceramic becomes a liquid the structure is of limited importance and the mobility of the individual ions become the primary means of transmitting energy. Because the mobility of the ions is an ever increasing function of temperature the thermal conductivity is also an ever increasing function of temperature for the stable ceramic into the gaseous state. Like the phase changes in the solid state, state changes will not produce a major change in the relatively low level of thermal conductivity that a ceramic displays. State changes may not be sufficiently large to appear as a distinctive break in the thermal conductivity curve. Glasses behave like liquids and converge with the solid state curve at the melting point.

Occasionally a ceramic will display thermal superconductivity at very low temperatures, usually well below the primary peak. Why this occurs has not been determined but it appears to have a similar mechanism to that of electrical superconductivity.

The type of primary peak a ceramic displays in thermal conductivity ranges from very gradual to very pointed. It can also be enormous relative to the subsequent level value. The shape of the peak tends to be dependent on the crystalline structure of the ceramic. A simple ceramic with a symmetric lattice will usually produce a sharp peak. The size of the peak can be linked to the basic conductivity of the metallic atom in the ceramic. The location of the peak can also be linked to the basic conductivity of the metallic atom in the ceramic. In both cases a conductive metallic atom will enhance the thermal conductivity of a ceramic such that the peak will occur later and be larger. These trends are depicted in Figure 3-4.

Multi-component thermal conductivity is highly dependent on the distribution of the phases. When a continuous phase exists it will dominate the amount of conductivity until very low concentrations. This will be true even when the non-continuous phase or phases are highly conductive and the continuous phase is not because the continuous phase becomes a barrier to be surmounted. If none of the phases are continuous than an averaging effect usually obtains. Porosity behaves like another phase and can have a noticeable effect on the thermal conductivity of a ceramic.

THERMAL SHOCK

Both glasses and ceramics are subject to thermal shock throughout their entire solid range. The low thermal conductivity of these materials does not allow them to quickly dissipate thermal differences as they arise during thermal cycling. A temperature gradient induces distortions due to the accompanying thermal expansion. It is the distortions that support cracking and lead to failure.

The greater the temperature gradient within the glass or ceramic, the more stress is induced and the more likely the material is to fail. In general it makes little difference if the material is being heated or cooled. Similarly, the rate of heating or cooling is not important in the absolute sense but rather as to the gradient it induces.

Properties of the material that reduce the thermal gradient or

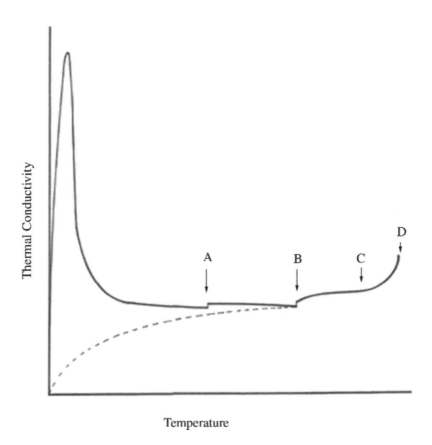

Figure 3-4 THERMAL CONDUCTIVITY OF A CERAMIC
This figure also shows a general schematic for a ceramic. The points A through D correspond to the same structural changes as described in Figures 3-1 through 3-3. The spike graphed here is of modest proportions and just below ambient. Usually the shift at point A would not be seen due to its small size relative to the error in measuring thermal conductivity at low temperatures. Similarly, the discontinuity at B might also go undetected. The dotted line corresponds to the glassy state behavior for the same composition as the crystal shown above it. On melting the glass would follow the solid line.

the stress associated with that gradient will improve the thermal shock resistance. Higher thermal conductivity is an especially effective means of reducing thermal shock. Composite materials are often assembled with this consideration. Very strong materials, which are able to tolerate high induced stress without failure, show good thermal shock resistance. Ductile materials, which can dissipate the thermally induced stress, are resistant to the extent of their ductility. Low density materials, especially glasses, often show fair thermal shock resistance because they have a structure that does not retain stresses.

At extremely low temperatures ceramics are so ridged that even crack propagation is slowed. Thus, this region will often be more resistant to thermal shock than a slightly higher temperature region. Beyond this extreme, the higher the temperature region in which cycling takes place the less resistant both glasses and ceramics tend to be to thermal shock. Higher temperature also aids other failure modes, which often contribute to the cracking.

Thermal fatigue is a closely related property to thermal shock. All of the factors that produce failure in thermal shock are involved with thermal fatigue. In addition, the glass or ceramic is subject to the crack propagation circumstances that make the material fail under conditions it could withstand for a few cycles. The cracking which the material experiences due to the thermal gradient produced are more easily propagated when the material is in the warmer part of the cycle. This acceleration in cracking leads to its quicker failure.

DIELECTRIC LOSS

Since most ceramics are insulators they are also good dielectrics. What loss they incur is mainly in the form of migration of ions. The more loosely bound ions of glass are more mobile inducing more dielectric loss.

Frequency of the imposed electric field also effects the amount of dielectric loss experienced. The higher the frequency the more vibration of the atoms. This increased vibrational energy is mainly dissipated in the form of heat. Increased frequency will increase the dielectric loss at all temperatures. This effect increases with increasing temperature. However, only at both high frequency and high temperature will the loss associated with vibrations approach that of the loss associated with ion migration.

All forms of dielectric loss increase moderately smoothly with increasing temperature until the Curie Temperature is approached. In this vicinity the dielectric loss peaks sharply and then decreases just as sharply. Even very small amounts of an additive can dramatically shift the Curie Temperature changing the loss peak. This approach to manipulation of the dielectric loss is especially useful when a high loss is desired.

If dielectric loss is to be minimized, then the most effective means is generally to reduce ion mobility. Ion mobility can be reduced by using the more tightly bound form of the crystalline ceramic. Additional reduction in loss would result by the removal of unbound and loosely bound additives. It can also be aided by eliminating defects, especially voids. Glass-ceramics are particularly useful as dielectrics because their processing makes them almost pore free.

It would seem that a reduction in grain boundaries within the ceramic would also help to reduce loss. This is true if a single crystal can be used. However, this is not always true of multi-crystalline ceramics due to magnetic considerations.

The dielectric constant follows the same trends as the dielectric loss because the same mechanisms are operable. A general plot of the temperature dependence of both the dielectric loss and the dielectric constant can be seen in Figure 3-5.

SUPERCONDUCTIVITY

Near absolute zero some ceramics display superconductivity. At such low temperatures the thermal vibrational energy of the atoms in the ceramic are very small and the atoms are frozen in position. If the ceramic is at an equilibrium structure of high purity and low defects, then the atoms are in very even spacings throughout the bulk of the ceramic. The ceramics that do display superconductivity have the metallic atoms so arrayed. The superconductivity comes about because electrons from the metallic atoms can be removed and transferred easily from metallic atom to metallic atom with an even wave-like motion. As the temperature increases, the vibrational energy of the metallic atoms becomes more and more disruptive of the symmetry. At the critical temperature some resistance to conduction of the electrons can be measured. Symmetry continues to decline with temperature increase with an accompanying decline in conductivity by electrons. Eventually electron conduction will become negligible and the ceramic will be an insulator unless hole conduction becomes significant.

During superconductivity, the movement of electrons through

the ceramic does not create heat as it would in a normal conductor. It is the heat production that usually causes the failure of conductors. Although superconducting electrons do not produce heat, they do produce a magnetic field. This field in turn produces internal stresses. When the internal stresses reach the yield point the ceramic fails in brittle fracture.

ELECTRICAL CONDUCTIVITY

Carbides are usually metallic conductors throughout their entire solid state temperature range; this is especially true if the metallic atom is a transition element. At low temperature they have high conductivity due to the ridged array of conducting metallic atoms. As the temperature increases the conductivity decreases but always retains a fairly large value. The warmer the ceramic the smaller the energy gap is for the carbon atoms until a point is reached when the gap is gone. At that point the carbon atoms become electron conductors and aid the metallic atoms. The higher the temperature the less affective the metallic atoms become at electron conduction but the more affective the carbon atoms become.

Other ceramics without carbon can also be electron conductors via their metallic atoms. The amount of electrons available for conduction is dependent on the ease with which a metal will give up an electron as well as the strength of the chemical bond that the metal has with its surrounding atoms. Metallic inclusions, like extra sodium atoms within a glass, are completely free to give up electrons but the number of such atoms will be limited.

Glass also contains many bound sodium atoms that will not readily give up an electron because of sharing with an oxygen atom. When a bound sodium atom dissociates from its neighboring oxygen atoms it contributes an electron that also leaves the oxygen atom free to give up an electron and become a positively charged ion. The electron that the oxygen gives up does not usually leave the area due to the positively charged metallic atom entrapment of it. The net result is a free electron and a positively charged oxygen.

Transition metals are able to give up electrons even if they are chemically bound because their unfilled subshells contain electrons what are almost at the conduction band in energy. Small fluctuations can provide the additional energy needed to pry these electrons loose. The result is a mobile electron with a stationary, positively charged metallic ion. The metal does not stay in this condition; it attracts and

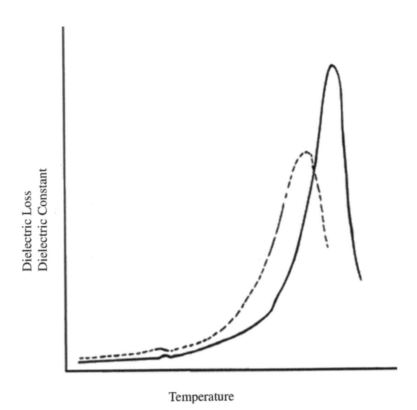

Figure 3-5 A GENERAL PLOT OF DIELECTRIC LOSS AND DIELECTRIC
CONSTANT

This graph shows the dielectric constant and the dielectric loss as a function of
temperature. The dielectric constant is indicated by the solid line and the dielectric
loss is indicated by the dashed line. The numerical values of these two properties
are very different, requiring two different scales for the same temperature scale.
What is obvious from the plot is how similar the relative values of these two properties are at any given temperature. The dielectric loss does slightly " lead" the
dielectric constant but the form and relative values are consistent. The major peak
is at the Curie temperature. Other minor peaks occur at temperatures associated
with structural changes.

81

traps available electrons as they move past. As a result there will be a small flow of electrons in the direction of the induced electronic field.

If the metallic electrons are not removable, the ceramic may still conduct electricity via ions. When the ceramic is a densely packed crystal at low temperature it is very difficult for ions to move. As the temperature is increased and vacancies are created some movement is possible. With increasing temperature, the vacancy number increases and the equilibrium structure becomes less dense. Non-crystalline phases like glasses or liquids greatly improve the mobility of ions. High temperature adds enough energy to the ceramic to permit both additional chemical bonds to break producing more ions and greater mobility of the ions.

ELECTRICAL CONDUCTIVITY IN MULTI-PHASE CERAMICS

Electrical conduction is very path dependent and charge carries use paths that accommodate their movement rather than a straight movement towards the conducting surface. The most common path for a charged ion to take is along the grain boundaries because the packing will be the lowest here. Electrons also tend to follow this path. As a result the charged particles will usually move through a multi-phase ceramic as readily as a single phase ceramic.

When the ceramic is made up of layers of different ceramics, which are aligned perpendicular to the applied electrical field, there will not be a phase boundary through which the particles can move. They will have to transverse the bulk of each ceramic in the order in which they are encountered. Thus, the charge carriers will zip through the conductive sections and seep through the less conductive areas. The result is the least conductive ceramic is the limiting material and its conduction will be that of the material as a whole. In calculating conduction only the thickness of the poor conductor need be considered rather than the entire thickness of the ceramic.

When a real ceramic conducts electricity it behaves as a multi-phase material even when its composition is entirely uniform. Voids act as a second phase with grain boundaries contributing to their continuous nature. Every grain within a ceramic has its own orientation; electrical conductivity varies with crystallographic orientation. In general the amount of variation with orientation is not large but there are some notable exceptions. Pyrolytic graphite in particular and graphite in general are very orientation sensitive.

Glass has an open structure that encourages ion mobility.

However, it does not have grain boundaries paths. As these effects may cancel each other, no generality concerning electrical conductivity of glass verses ceramic of the same composition can be made.

ELECTRICAL CONDUCTIVITY OF GLASS-CERAMICS

Both glasses and glass-ceramics use ions as the principle means of conduction; they both show increasing conductivity with increasing temperature. However, a glass-ceramic usually has significantly lower electrical conductivity than the glass from which it forms. This is due to the reduction of the mobility of ions " with the crystalline phase. However, if the crystalline phase does not incorporate the mobile ions when it forms then these ions tend to coalesce in the glass phase because of the additional room for them there. When this happens the mobile carriers are concentrated in the less dense continuous phase and the glass-ceramic will actually have a higher electrical conductivity than its parent glass.

When a lower electrical conductivity is desired in either a glass or a glass-ceramic it can be accomplished by reducing the number of mobile carriers, which is usually the sodium. In some processing it is possible to replace the mobile sodium ions with less mobile lithium ions. If metallic ions cannot be removed in sufficient quantity to reach the desired level than the ions can be blocked in their movement by larger ions. Both lead and barium are used for this purpose.

MAGNETIC PROPERTIES

Any particle that carries a charge and is in motion produces a magnetic field. The strength of the field is directly proportional to the mass, motion and charge of the particle. A magnetic field will not be effective if it is in the presence of another field or fields that negates it by being in a different direction. Even a very small field can become important when it is joined to other fields that are orientated in the same direction.

A proton in the nucleus of an atom has a large mass in comparison to an electron but it has very little motion. Since protons have alternating spins even numbers of protons in a nucleus result in no net magnetic field because the pairs cancel each other. Only with odd numbers of protons in the nucleus will there be nuclear magnetic resonance. This will still be only about one millionth as large an effect as from the electrons surrounding the nucleus. However, there are millions of nuclei within a material and not much of an impetus to rotate the nucleus in interfering ways. Proton produced magnetic fields have found use in the medical area as well as an experimental tool for exploring the subatomic nature of ceramics.

The main potential for development of a magnetic field lies with the electrons of a ceramic. The outermost electrons are usually involved in chemical bonding. By its nature, the chemical bond requires that the electrons be of the opposite spin. Thus, any pair of chemically bound electrons will not produce a net magnetic field. Similarly, subshells within the bound atom usually have paired electrons. If unfilled electron subshells are present then they will contribute what appreciable magnetic field a ceramic will have. Only transition series metals have unfilled subshells; there are many of them, and their presence in ceramics is extensive. The strongest of these is iron.

If all of the available unpaired electrons within a single magnetic domain are aligned in the same direction than the ceramic is ferromagnetic. If some of the unpaired electrons are oriented in the opposite direction to the overall magnetic field than the ceramic is ferromagnetic. If the opposite spins exactly equal the field direction such that no magnetic field results than the ceramic is anti-ferromagnetic. See Figure 3-6 for diagrams.

A common source of a magnetic field is the flow of electrons during electrical conduction. Because ceramics are poor electrical conductors they do not contribute significantly to this area of magnetic application.

The rigid crystalline structure of solid ceramics produces a far more ordered arrangement for the unpaired subshells of the transition elements within the ceramic than either polymers or metals. This allows an induced magnetic field to be maintained easily. The lower the temperature the more rigid the ceramic structure and the better the alignment can be. Increasing temperature causes the thermal vibration to distort the crystal. Eventually there will not be enough order to maintain the magnetic field strength and the ceramic will become paramagnetic at the Curie temperature. Paramagnetism is also displayed by ceramic glasses because they do not possess the highly ordered arrangement of the crystalline state.

The substitution within a ceramic of one magnetic transition metal atom by another atom will affect the value of the net magnetic field, although not usually the direction. A fair amount of substitution must take place before much change in the net magnetic field is real-

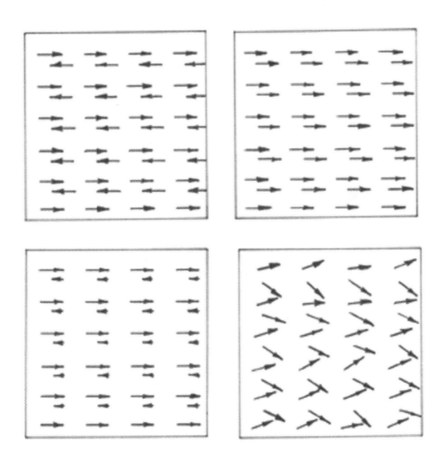

Figure 3-6 MAGNETIC FIELDS IN CERAMICS
The vectors associated with the magnetic fields produced by individual atoms within a crystalline ceramic are shown as arrows in this figure. The upper left block shows an anti-ferromagnetic ceramic. For the vectors to exactly match in both size and alternating direction usually requires that the entire field be produced by the same type of atom. The reason they are pointing in different directions is associated with their bonding directions within the crystal. Directly below anti-ferromagnetism is the block depicting ferromagnetism. This is basically the same as anti-ferromagnetism but produced by different types of atoms and the size of the vector is associated with the atom type. The upper right block depicts the vectors that comprise a ferromagnetic ceramic within a single domain. The lower right block shows paramagnetism within a ceramic. This type of ceramic is ferromag-netic below the Curie temperature where the crystal is in better alignment.

ized because the magnetic field is relatively composition insensitive. However, any change in the structure will greatly alter the direction of every magnetic vector in the entire array. Thus, magnetic ceramics are considered structure sensitive.

A strong externally applied magnetic field will induce the individual magnetic vectors towards alignment with the field. This can only be done by introducing a distortion to the chemical bond so that the atoms can rotate relative to the field. The distortion to the bond then introduces stress into the ceramic. The applied field will result in a change in the magnetic domain pattern of the ceramic to reflect the size and direction of the applied field as much as possible within the constrains of a real material.

Domain walls are contained within individual grains unless the mismatch between grains is very slight. Besides grain boundaries limiting the shape and size of the domains, non-magnetic inclusions and pores can also restrict them. The most effective of these will be the inclusion or pore that matches in size the width of the domain wall. If it is smaller, it will not be large enough to hold the domain wall. If it is larger it can act as a radiation site for the generation of domains which can move away. The width of the domain wall is characteristic of the ceramic and can be measured by viewing the polished surface of the ceramic with polarized light. It is usually about 10^{-5} cm.

PIEZOELECTRIC PROPERTIES

If a crystalline ceramic develops an electrical charge proportional to an applied mechanical stress then it is piezoelectric. The mechanical stress the piezoelectric ceramic responds to can be a directly applied force or it can be the result of indirectly applied stress such as non-uniform thermal expansion. The electrical charge that is produced is the result of established dipoles in the crystal when the structure does not have a center of symmetry in the direction of the applied field. It is this lack of symmetry that allows the polarization. Most evidence suggests that atom motion is the mechanism of polarization.

When a piezoelectric ceramic responds to a uniform thermal expansion such that a dipole motion is induced, than this ceramic is both piezoelectric and a member of the subset pyroelectric. If the direction of the dipoles can be reversed, by the application of an electrical field, then the ceramic is both piezoelectric and a member of the subset ferroelectric. A ceramic that is pyroelectric, wherein the polarization can be reversed via an applied mechanical stress, is ferroelastic.

Spontaneous, reversible polarization was initially considered directly analogous to ferromagnetism; thus the terminology was borrowed for ferroelectricity. Ferroelectric ceramics do show similar trends to ferromagnetic ceramics but the Curie Temperature of the ferroelectric ceramic is not the same as the Curie Temperature of the ferromagnetic ceramic. Also, ferromagnetic ceramics are not necessarily ferroelectric and vice versa.

Ferroelectric behavior is mainly a low temperature phenomenon with the highest known Curie Temperature being 675° C. Curie Temperature is moderately increased by an increase in pressure and dramatically decreased by the presence of impurities. At the Curie Temperature the dielectric constant suffers a notable spike in its otherwise rather stable value. This is illustrated in Figure 3-7. Just above the Curie temperature the ceramic may be subject to significant electrostriction.

OPTICAL PROPERTIES

In ceramics both the outer and the inner electrons are sufficiently tightly bound to the parent atom that they are in little danger of being dislodged from ordinary light sources. Thus, no electrons are just floating freely within the ceramic and the entire atom responds to incident light. Each atom absorbs energy at its own characteristic wavelength range associated both with its own nature and the overall composition of the material in which it is located. The wavelength range varies slightly with both temperature and pressure. The amount of "background" absorption by each atom is also variable over the whole spectrum of wavelengths. As it happens many of the atoms that commonly compose ceramics do not have energy absorption peaks in the visible wavelength range and are thus able to transmit most of the incident light through the ceramic.

If a thin section of defect free ceramic is made it will often be very transparent because light passes through it without hitting any particles. As the number and type of defects are increased, the transparency is correspondingly decreased because the regularity of the structure is disrupted by defects that can trap and hold light. If light is absorbed by an atom within the ceramic, it is usually re-emitted causing the reflection of the light from the ceramic. The light that is initially absorbed by the ceramic and not reflected is retained as heat energy.

Impurities that are not chemically bound to the ceramic are particularly effective in absorbing light because they have free electrons available. If an impurity atom absorbs light and then only partially reflects it, with a change in the wavelength characteristic of that impurity, it is possible to create a color center. The color source is the impurity, but due to the spreading of the light the entire ceramic takes on the appearance of that color. The intensity of the color is directly proportional to the number of color centers. If multiple color centers of different types are present than the net color will be the mixture of these colors.

The perfection of a ceramic decreases with increasing temperature and decreasing pressure. It would follow that the transparency of a ceramic would be effected by temperature and pressure in the same way that crystalline perfection is effected. This is generally true. However, it is possible for even relatively unstructured ceramics to be transparent if they are not too thick and are composed of atoms that do not absorb readily in the visible range. This seeming contradiction comes about because two different aspects of light absorption are in effect. In the first consideration, the perfection of the crystal insures that the atoms are in their expected positions and the light will be able to pass between the atoms without striking one. In the less structured ceramic this is also possible if there are few atoms to hit. Thus, with a thin section or a relatively open structure, light can pass through without striking many atoms. The most important of these ceramics is silicon based glass. A ceramic that is both highly structured and widely spaced will give a very good path for light to pass through. Thus, a crystalline ceramic with low defects in the loosely packed direction will have excellent transparency for thin sections at some part of the energy spectrum. Figure 3-8 shows a general absorption curve for a ceramic.

This same transparency to visible light of many ceramics is also shown in other areas of the energy wave spectrum. The area that has caused the most interest in recent years has been the transparency to microwaves.

PHOTOSENSITIVE GLASS:
Glass is not as completely bound chemically as a crystalline

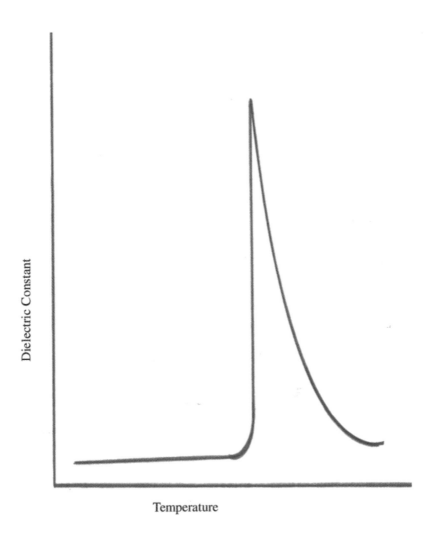

Figure 3-7 THE DIELECTRIC CONSTANT OF A FERROELECTRIC CERAM-
IC
This graph shows the general behavior of a ferroelectric ceramic with the dielectric
constant plotted as a function of increasing temperature. In most ceramics the
dielectric constant increases very slowly with increasing temperature until close to
the Curie Temperature. Well above the Curie Temperature the dielectric constant
remains at a moderate value. It is only in the vicinity of the Curie Temperature
that major changes in the dielectric constant occur. Ceramics generally have one
Curie Temperature above which the ceramic is no longer ferroelectric. However,
multiple Curie Temperatures are possible with each one affecting the dielectric
constant.

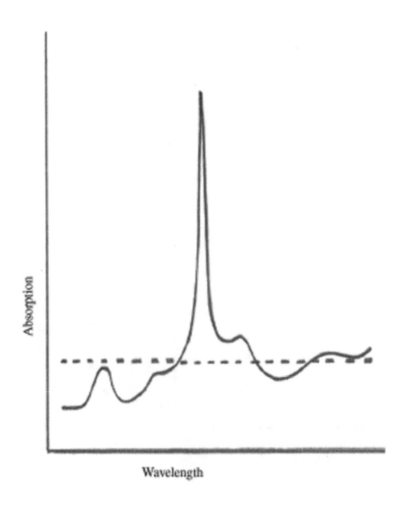

Figure 3-8 ABSORPTION OF INCIDENT ENERGY BY A CERAMIC
This graph is a general schematic of a ceramic absorbing energy as a function of
wavelength. For relatively narrow ranges the background absorption can be
approximated by a single value which is shown as a dotted line. When this
approach is used only the high absorption peaks need be noted. With increasing
temperature or decreasing pressure, the entire curve is shifted upward on the
absorption axis with little change in curve shape.

ceramic. In common glass the oxygen atom is only connected to the whole by a single bond whereas it is capable of two bonds. The single bonded oxygens attract and hold free electrons, which creates an electrical hole. It is this stable electron/hole which acts as a color center in the glass.

Electrons can be liberated by an energy source outside the glass such as light. The number of electrons generated is dependent on both the intensity and the duration of exposure. As these electrons are generated they will tend to attach themselves to the single bonded oxygens until all of the single bonded oxygen atoms have a free electron associated with them. During this process the photosensitivity of the glass will increase as a new color center is generated each time an electron is attached to an oxygen.

Besides the color centers generated by the oxygen/electron pair, the glass may contain color centers due to impurity atoms. It is the color centers, produced by the impurity atoms, which usually determine the visible color of the glass. However, the depth of color and the distribution of color throughout the glass are more dependent on the oxygen/electron pair number and location than on the impurity atom number and location because of the much greater number of oxygen/electron pairs. Once the glass has become colored due to this process it tends to remain so even when the energy source is removed.

PHOTOCHROMIC GLASS:

Silver halide atoms have been deliberately added to glass to serve as color centers. This compound and other similar compounds react to visible light by chemically changing in a reversible way. By pairing a particular glass with a particular light sensitive compound, glass composites of a highly responsive nature can be produced.

This process may also make use of the singularly bonded oxygen atoms in the glass structure to hold the dissociated compounds. However, more success has been achieved in the straightforward use of photosensitive ceramics interspersed in fine particle form throughout the glass host.

TENSILE PROPERTIES

TENSILE TEST

The tensile test is the most commonly used from of mechanical testing. In general it is done on multicrystalline or glassy phase ceramics of commercial interest. Other more difficult and subtle forms of testing are reserved for single crystal and experimental materials. Testing conditions and techniques do vary and are not as readily comparable as it might appear.

The general procedure for the tensile test is to apply a load to the ceramic in a simple manner and continue its application until the ceramic breaks. Assumptions inherent in the test are atmospheric pressure and ambient temperature. These two conditions should not be varied without being noted. While the speed of loading as well as the gripping conditions are specified by the American Standard Testing Method these are not always practical for all ceramics. Variations to the simple loading and gripping procedures specified by ASTM include a more gradual loading for particularly load sensitive ceramics and cementing the grips for particularly crush sensitive ceramics. Any such change in procedure should be noted and must be considered in comparing results.

The surface condition of the specimen is an important variable and often has dramatic effects on the test results. Some ceramics polish readily and are given a smooth surface prior to testing. Others are difficult to polish and are tested in the as formed condition. Still others must be cut to the test dimensions but are not capable of being polished. Obviously the surface roughness of these various samples will vary enormously. Ceramics are very sensitive to surface cracks or cuts which propagate during tensile testing. Test results often reflect the surface conditions more readily than the actual strength of the ceramic.

COMPRESSION TEST

Like the tensile test, the compression test applies a load to the ceramic according to specifications provided by the ASTM. The primary difference in the two tests is the direction of loading. Compression tests also vary from tensile tests in specific conditions of shape and end loading but the results are roughly comparable. Compression tests are particularly useful for ceramics due to the extensive use of ceramics under this loading condition. In general a ceramic will be stronger in compression and less sensitive to surface and loading conditions than the same ceramic in tension.

ELASTIC BEHAVIOR OF BRITTLE CERAMICS

Whenever a force is applied to a brittle ceramic the initial response will be a distortion of the bonding of the overall matrix. Once the force is removed, the bonding shape will return to its original form. See Figure 4-1 for an illustration. It is only bond distortion that is truly reversible and unequivocally elastic in nature. Brittle ceramics are limited in their ability to distort the chemical bonding that holds them together and therefore are limited in their true elastic behavior. Any shifting of atoms or breaking of bonds to allow additional distortion of the material is of a non-elastic type.

Brittle ceramics begin to produce small microcracks on the atomic scale even before the applied force has reached the elastic limit of bond distortion. The amount and the direction of the microcracking depends directly on the amount and direction of the load. The response of a brittle ceramic under increasingly higher loads can be seen in Figures 4-2 through 4-7. In all cases these ceramics are considered to be displaying elastic response to the applied load although the cracking becomes ever more pronounced.

ELASTIC BEHAVIOR OF DUCTILE CERAMICS

Ceramics display brittle behavior at very low temperatures but as they are heated they become increasingly ductile. The difference is the sensitivity of the ceramic to the propagation of cracks both on the microscopic and macroscopic level. In the brittle range even the smallest void can serve as an origin for a crack. How easily the crack can propagate is dependent on the crack orientation relative to the applied load, the general shape of the crack and the bonding strength of the ceramic in the direction of crack propagation.

On heating ceramics are more capable of ductile response to the stress concentration of a propagating crack due to localized bond distortion and realignment. The result is less cracking within the ceramic and improved strength. However, as the temperature continues to be increased, the gains in strength due to reduced cracking are more than compensated by the losses due to ever decreasing bonding

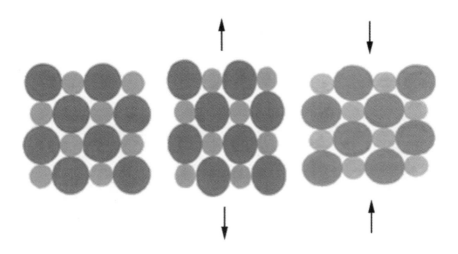

Figure 4-1 ELASTIC RESPONSE OF A CERAMIC
As in previous figures, the lighter circles depict metallic ions and the darker circles
depict non-metallic ions. The figure on the left shows a ceramic in the unstressed
condition. The figure in the middle shows the same ceramic with an applied tensile
force. The figure on the right shows the ceramic with an applied compressive force.
The forces are sufficiently great to distort the shape of the chemical bonds but not so
strong as to break the bonds. This is true elastic response, which is completely
reversible.

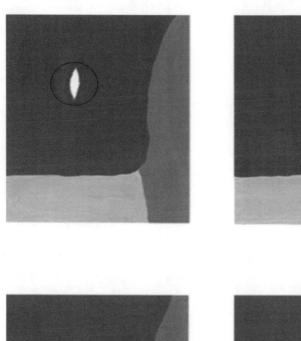

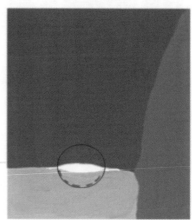

Figure 4-2 BEHAVIOR OF A VOID UNDER TENSION

A representation of four microstructures is shown in this figure. In each case the grains are depicted as gray areas of differing shades depending on their relative orientation. The grain boundaries are the intersections of these areas of different shade. Crack areas are highlighted by the elipses The upper left shows a void within the bulk of a grain that is in the same direction as the tensile force. This void will be relatively unaffected by the tensile force applied. The upper right shows a similar void in the grain boundary. The response to the tensile force by the void will also be minimal. The lower left shows a void within the bulk of the grain but perpendicular to the direction of the applied tensile force. This void will tend to open and propagate cracks. Similarly, the void along the grain boundary of the lower right microstructure will also easily open and crack along the grain boundary due to the applied tensile force.

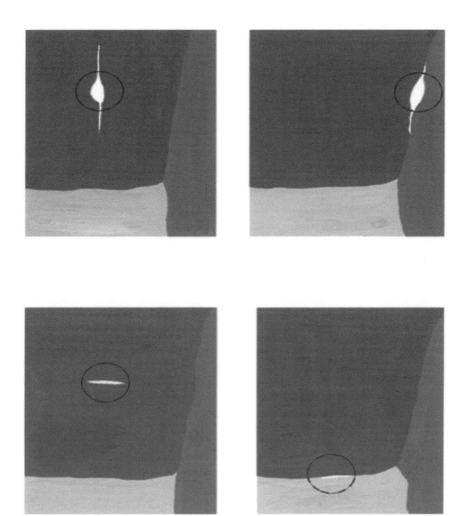

Figure 4-3 THE BEHAVIOR OF VOIDS UNDER COMPRESSION
As in Figure 4-2, this figure depicts microstructures with a void. In this case the applied force is now in compression; elipses are also used in this illustration. In the upper left microstructure the void was relatively unaffected by the tensile force but it shows both a bulging and cracking under compression. A similar response is shown in the upper right microstructure. In tension it was stable but in compression it both originates and enlarges cracks. By contrast, the lower left microstructure shows the voids constricting under compression. The same is true of the lower right microstructure.

97

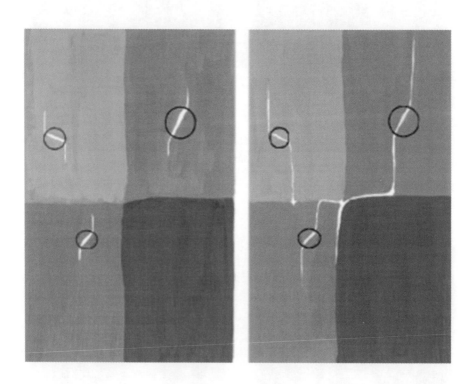

Figure 4-4 PROPAGATION OF CRACKS UNDER TENSION
The schematic on the left shows voids beginning to crack under a tensile load. The cracks are in the direction of the load and usually at the ends of the voids. As the cracks develop, as shown in the schematic on the right, they encounter grain boundaries and follow these in preference to the bulk of the material. An exception occurs when the rate of the applied load is greater then the speed of the crack propagation. Usually only explosive loads produce cracks that do not follow the grain boundaries in brittle ceramics.

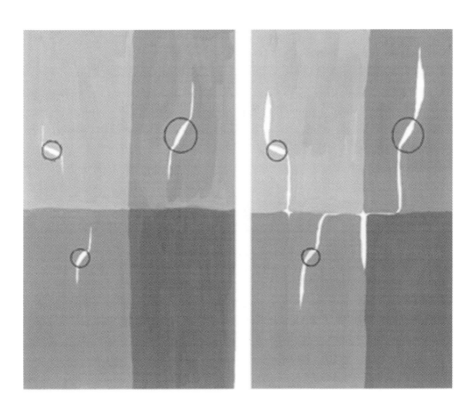

Figure 4-5 PROPAGATION OF CRACKS UNDER COMPRESSION
This figure is very similar to Figure 4-4. The major difference is the width of the cracks as they propagate due to the compressive load. Cracks that are parallel to the compressive load broaden. Cracks which travel across the grain boundaries perpendicular to the compressive load do not broaden.

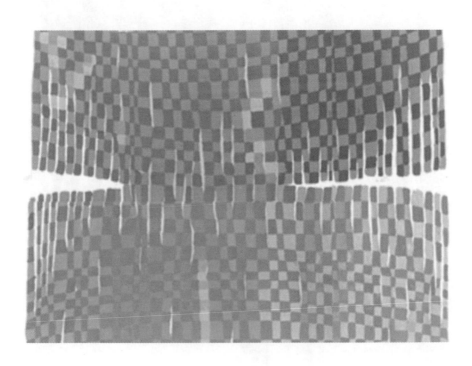

Figure 4-6 CROSS-SECTION OF A BRITTLE CERAMIC IN TENSION
The previous four figures show a magnified area of a ceramic, whereas the next four figures show the center section in cross-section at approximately full size. The checkerboard pattern is a simplification of a real grain structure. In general brittle ceramics under tension produce a sharp break in the center of the material.

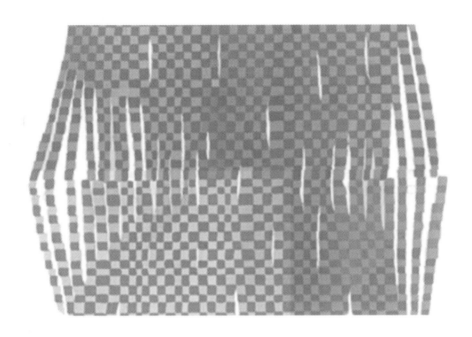

Figure 4-7 CROSS-SECTION OF A BRITTLE CERAMIC IN
COMPRESSION
Like Figure 4-6, compression in a brittle ceramic usually produces a break through
the center of the ceramic. However, the compressive load broadens the cracks such
that splintering occurs around the sides of the ceramic. The center tends to produce
breaking into individual grains or powdering.

101

strength and ever increasing vacancies. Because ductile ceramics are not nearly as sensitive to microcracking, they can display a relatively large range of elastic behavior, which is moderately reversible. Some propagation of cracks is to be expected even within this range.

ELASTIC BEHAVIOR OF GLASS

Glasses have a less ridged structure than a crystalline solid and are more easily deformed elastically. The individual bonds are able to both elongate and twist relative to the applied force. It is this twisting aspect that allows glass additional extension as compared to a ceramic of the same composition. In addition, glasses are less subject to cracks and crack propagation in the elastic region.

The amount of twisting possible is directly related to the structure of the glass. A highly bridged glass with be less able to move in response to the applied load and the elasticity will only be slightly greater then the corresponding ceramic. It is in glasses with little bridging that relatively high elasticity is displayed.

PLASTIC BEHAVIOR OF CERAMIC

Ceramics are not noted for their ductility, but there are specific conditions in which it will be displayed. Some ductility can be seen around cracks at most temperatures. This localized deformation may retard the crack and add to the strength of the ceramic. However, it does little to add to the elongation of the ceramic. General slippage of one grain over another also contributes to ductile behavior, but grain slippage is limited by grain boundary intersections, which serve to pin the grains.

Plastic deformation in a ceramic within a grain is very complex due to three-dimensional chemical bonding. Slip of one plane of atoms over another requires breaking and reforming of chemical bonds such that the final result is a permanent displacement of the atoms. Each bond within a plane of slip must not only break and reform with a further neighbor, the new bond must be with the appropriate type of atom and in an alignment that allows bonding. Furthermore the bonding electron orbitals of s-s, s-p or p-p must be maintained in the rebonding.

Localized movement is difficult because ceramics are bonded such that breaking a single bond does not free an individual atom. There is an overall sharing of electrons that pervades through the structure of each grain. Massive movement within even the less densely packed planes is not energetically favorable at lower temperature where the bonds are strong. Only very weak bonding directions and very high temperature can support plastic flow. Plastic flow is illustrated in Figures 4-8 and 4-9. A comparison of brittle verses ductile strength is shown in Figures 4-10.

As temperature increases, the vibrational energy increases, making the chances of any one atom moving in a specific direction greater. Increased temperature also brings increased vacancies, which aid in allowing the lattice to relax into their space. When a force is applied it prejudices the random movement of individual atoms in the direction most able to respond to the force. This direction will only be approximately the same as the applied force due to the necessity for bonding to maintain specific relationships between elements. The crystal will slip along specific directions that allow for appropriate rebonding. Between high temperature and high applied force, the vibrational energy of the atoms could be sufficient to allow momentary breaking and reforming of an entire plane. However, if the lattice needs to move more than one atom over to rebond, the probability of alignment time for both the movement and the rebonding decreases with the atomic distance needed to transverse. Complex ceramics do not display true ductility for this reason.

A ceramic generally deforms in a manner similar to the dominant metal from which it is formed. A simple carbide like tungsten carbide will have the main attributes of tungsten at a similar temperature. This suggests that the deformation is primarily metallic in nature using the same slip mechanisms to produce ductile slide. The analogy may be pursued to predict the ductile behavior of yet untested ceramics provided only like ceramics are compared.

Like ceramics will display similar ductility because they have similar bonding. For example, oxides consist entirely of the s form of bond. Both the oxygen and the metal bond only in this way. This bonding is not as angularly dependent as the p form of bond although there seems to be some preferred orientations. A lattice made up entirely of the s form of bond is easier to align for rebonding than a mixed bonding configuration because there is no problem of matching bond forms. However, the s-s form is the strongest of the possibilities and requires the most energy to break. As a result, oxides tend to display ductility at a higher temperature and the increase in ductility with increasing temperature will be greater then most other ceramics.

In contrast to oxides, carbides combine both s and p forms of

103

Figure 4-8 CROSS-SECTION OF A DUCTILE CERAMIC IN TENSION
Ideally a ductile ceramic will elongate uniformly throughout. However, due to loading conditions, the center of the ceramic will usually experience a greater load than
the rest of the length causing the center to neck. Once necking has begun, subsequent plastic deformation will be in this region. Eventually the ceramic will pull
apart. This figure is primarily to depict the distortions the grains experience under
a tensile load. The original square grains are most elongated at the surface of the
necked ceramic.

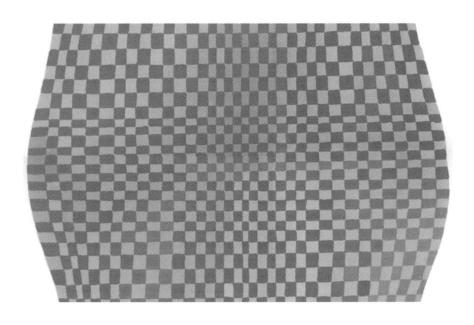

Figure 4-9 CROSS-SECTION OF A DUCTILE CERAMIC IN COMPRESSION
End conditions usually effect ceramics so that they do not compress uniformly. The center section experiences barreling where the outside grains are the most distorted relative to the center grains.

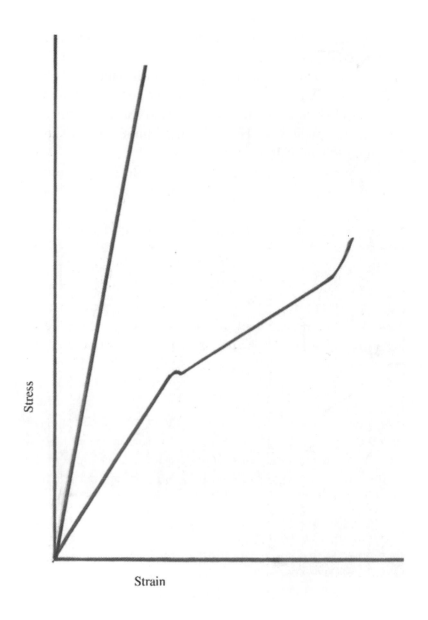

Stress

Strain

Figure 4-10 GRAPH OF TENSILE RESULTS
This is a plot of tensile test results for a brittle ceramic, on the left, and a ductile ceramic, on the right. The brittle ceramic appears to be more elastic in behavior than it actually is. During the test cracking will commence at all but the lowest force levels. The ductile ceramic usually shows an initial straight region corresponding to moderate elasticity. Next is a change in slope segment that corresponds to the onset of necking. The second straight line segment is the ductile region. The final segment is markedly non-linear to failure. Extensive cracking takes place in this segment of the test. A graph of compression results is rarely produced.

bonding orbitals. Carbon generally forms s-p hybrid bonds with metals creating a uniform, directional bond. The directionality of the bond, combined with the necessity to maintain the lattice structure, makes movement across carbon-metal planes difficult. The s-p hybrid form of bond is weaker than either the s-s or non-hybrid s-p form of bond. The s-p hybrid bond allows carbides to show ductility at lower than average temperatures. However, the limitations on slip planes tend to inhibit the amount of ductility.

Graphite is an example of a ceramic with highly directional p-p form of bonding, which is illustrated in Figure 4-12. Because the very long p-p form of bonds are also very weak, these are easily broken. Alignment within the plane of the p-p form of bonds is consistent, allowing both breaking and reforming easily in that one direction. As a result, graphite is ductile when the crystallographic orientation is random, because at least some of the grains will be favorably orientated for slip. Pyrolytic graphite is an oriented structure and only ductile in the direction perpendicular to the p-p form of bonding plane.

Some ductility can be shown by plate like structures. Plates are loosely bound to each other, often only by hydrogen bonds. This loose bonding allows for slipping between the plates although there will be no slipping within individual plates. If the ceramic is polycrystalline with a random orientation to the crystals, than the ceramic will approximate ductile response. If the ceramic is not polycrystalline or the orientation is not random, then the ceramic will display ductile response only when the applied force is perpendicular to the plate surface.

Plastic deformation relative to an applied load does not always increase with increasing temperature. An initial increase in plastic deformation of a ceramic is often offset by a subsequent decrease in elastic deformation. The result is a peak in strength prior to the melting point, which is illustrated in Figure 4-11.

PLASTIC BEHAVIOR OF GLASS

In the most highly bridged structures, glass is not more plastic than a ceramic of the same composition. The resistance to plasticity comes about both from the number of bonds and from their orientation. Bridges can be preferentially located in specific directions that will adversely effect plasticity in just that direction. A highly bridged glass may be considered to more closely resemble a highly cross-linked polymer than a crystal.

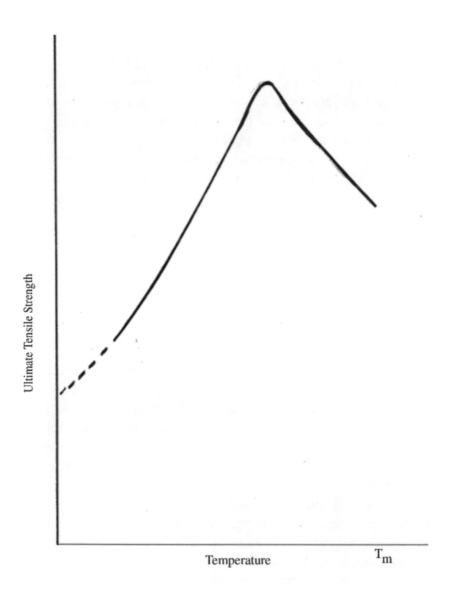

Figure 4-11 TEMPERATURE VERSES TENSILE STRENGTH OF A DUCTILE
CERAMIC

When tensile data like that in Figure 4-11 is compiled as a function of temperature
a graph similar to this one results. The ultimate strength, the strength at failure, is
plotted verses temperature regardless of the mode of failure. Even for the most ductile of ceramics, the lower temperature tensile responses is of a brittle nature. As
ductility increases with temperature so does strength until a maximum is reached.
Occasionally this maximum is at the melting point but usually it is decidedly below
it.

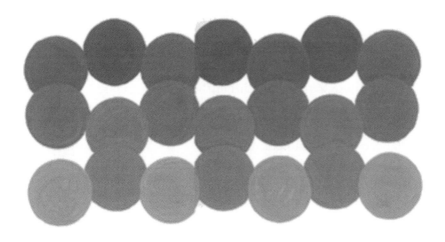

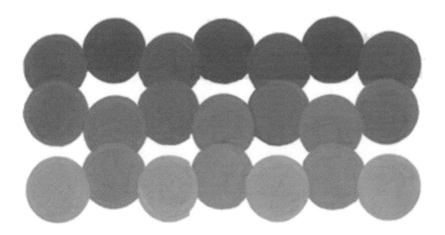

Figure 4-12 THE STRUCTURE OF GRAPHITE
The gray balls depict carbon atoms within a graphite structure. The first row is the lightest with subsequent rows becoming progressively darker. The (100) plane is shown at an angle to the plane of the front, which is the (001) plane. Obviously the (100) plane is very tightly packed and has a correspondingly strong bond. The (001) is very widely spaced and has a much weaker bond. It is this easy breaking of bonds in the (100) planes which allows sliding to a new position and rebonding, giving graphite its lubricating qualities.

When a glass is relatively free of bridges, it can respond like an entwined polymer. Various segments of the glass can move as a unit past other segments. This type of plastic deformation is usually possible in all directions.

Glass becomes more ductile with increasing temperature primarily because bonds break on heating. As a result, glass is heated to form into desired shapes, which can be quite intricate. After being formed at a high temperature and cooled to a lower temperature, the glass will regain stiffness while retaining the shape into which it was formed at the higher temperature.

TENSILE BEHAVIOR OF GLASS-CERAMICS

Glass-ceramics usually have sufficiently good bonding between the two materials that the composite behaves more like a homogeneous material. While the ceramic part of the material is able to carry a larger load then the glass, it is more subject to crack propagation. If the glass is processed such that the surface is in compression both at the edge of the material and at the interface of the glass and the ceramic, the glass can greatly mitigate the cracking. When the glass-ceramic is utilized in compression, this condition is maintained. However, when the glass-ceramic is put in tension the protective compressive layer of the glass can be negated. The cracking radiating from the surface will be slowed by the stronger ceramic segments, but it will still fail quickly once this state is reached.

THE EFFECT OF PRESSURE ON TENSILE PROPERTIES

In other materials the effect of pressure is fairly straightforward. Uniform pressure decreases the mobility of the material and thus increases its strength and decreases its deformation. In general this is what happens to ceramics but it does not hold well enough to risk a design based on this assumption. The exceptions in ceramics can prove both spectacular and disastrous.

Polymers and metals are both sufficiently ductile that small cracks will be blunted at their tip and not propagate easily under increased pressure. In ceramics there is not usually sufficient ductility to render these cracks ineffective. In addition, the added energy of the pressure can actually serve to propagate the cracks which would not be activated under atmospheric conditions. If uniform pressure will serve to aid or abate cracking in a particular ceramic is very difficult to determine without experimentation.

Hydraulic pressure is of special concern because many ceramics and glasses are sensitive to water. Water may be absorbed into the structure both physically and chemically. Additionally water may react with the material and alter its structure and strength. In general, low temperature and high pressure result in an increased strength in the material over its normal strength. As the temperature increases the reactivity of the water increases which will negate the effect of the pressure. Around ambient it can be expected that the corrosive nature of the water will have noticeably reduced the strength of the material. Glasses are even more likely to suffer than ceramics due to their more open structure, which allows the penetration and chemical reaction of water.

THE EFFECT OF IMPACT LOADING ON TENSILE PROPERTIES

Sudden impact on any material can lead to unexpected failure despite the impact force being below the tensile strength. This is particularly true of instantaneous forces such as those transmitted electronically; the usual mechanical impacts are not generally as devastating. If the material has insufficient response time to deform, it will shatter.

Within the more usual conditions of mechanical impact, ceramics still tend to shatter. This is due both to the general lack of ductility of the ceramic and to pre-existing cracks. Repeated impact loading is always a source of general cracking in a ceramic whereas a more ductile material would only sustain surface damage. Impact loading is to be avoided in ceramics unless the ceramic is solidly within a ductile region of its behavior range.

THE EFFECT OF CORROSION ON TENSILE PROPERTIES

Ceramics are very resistant to corrosion under non-stressed conditions. However, when a force is applied chemical bonds within the ceramic break and the site of these broken bonds becomes chemically active. The bonds may reform within the ceramic or they may form with available free atoms. When the new bonds are formed with available atoms the ceramic is considered corroded. The most common corrosion agent is oxygen in the air. Other atoms, such as impurities, can produce corrosion products as well. The negative effect of the corrosion reaction is two-fold. First, it destroys a chemical bond within the ceramic thus weakening it. Second, it produces a product that does not physically fit within the structure of the ceramic. If the product of the corrosion is smaller than the original site then the structure relaxes into the vacated space. Usually the corrosion product is larger than the original site and this causes additional stress on the ceramic due to crowding. The added stress will in turn encourage more bond breakage and further corrosion.

By its nature, corrosion is detrimental to the strength of the ceramic. The combination of corrosion and applied force accelerates the deterioration of the strength of the ceramic because the applied force both aids in the corrosion and pre-disposes it to follow a specific direction relative to the applied force. When the applied force is tensile the corrosion tends to proceed perpendicular to it because the most active sites are at the tips of the expanding cracks. Figure 4-13 illustrates this. When the applied force is compressive the corrosion tends to proceed along the direction of the force for the same reasons as the tensile response. Figure 4-14 illustrates this.

THE EFFECT OF DENSITY ON TENSILE PROPERTIES

The higher the density of a pure ceramic or glass, the stronger it will be in tension. Higher density indicates fewer vacancies and other types of voids. However, if the material is not pure, the higher density may result from heavy impurity atoms or corrosion products filling the voids. These will usually not contribute to the tensile strength of the material. Figure 4-15 shows the general relationship between density and tensile strength.

The behavior displayed in compression for pure ceramics or glasses is similar to that of tension. Impure materials of high density will usually be stronger in compression than would be expected. The high-density impurities contribute to the compressive strength both by their own high strength and by blunting cracks.

The lower the temperature the lower the equilibrium number of vacancies in the material. Similarly, the lower the temperature the shorter the bond length. Both contribute to the overall density and strength of the material. Structural phase changes with decreasing temperature also favor a more dense material. Figure 4-16 shows two graphs. The upper is the change in density with temperature of a ceramic. The lower is the corresponding change in tensile strength with temperature. Similar behavior can be expected for compression.

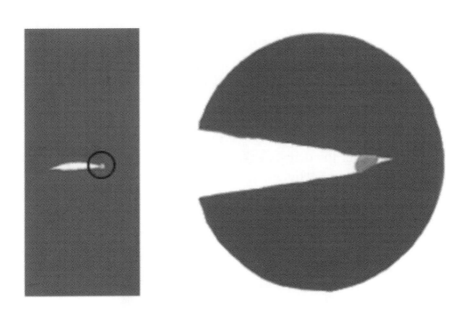

Figure 4-13 CORROSION IN CONJUNCTION WITH A TENSILE LOAD
The diagram to the left depicts a ceramic with one crack under a tensile load. The end of the crack is an area that preferentially supports corrosion. This corrosion product is shown as the light gray spot. To better see the corrosion product the circled section has been enlarged and is to the right of the specimen.

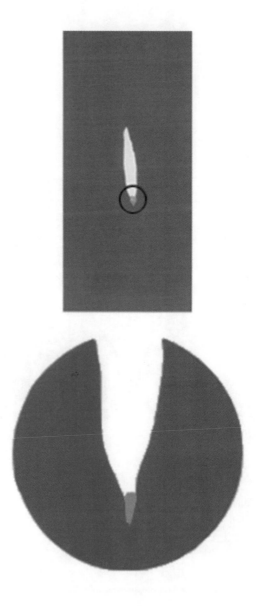

Figure 4-14 CORROSION IN CONJUNCTION WITH A COMPRESSIVE LOAD
As in Figure 4-13, the ceramic is depicted on the left with an enlargement of the circled area to the right. The difference between these figures is one of crack orientation relative to the direction of applied load rather than a distinction in the corrosion
mechanism.

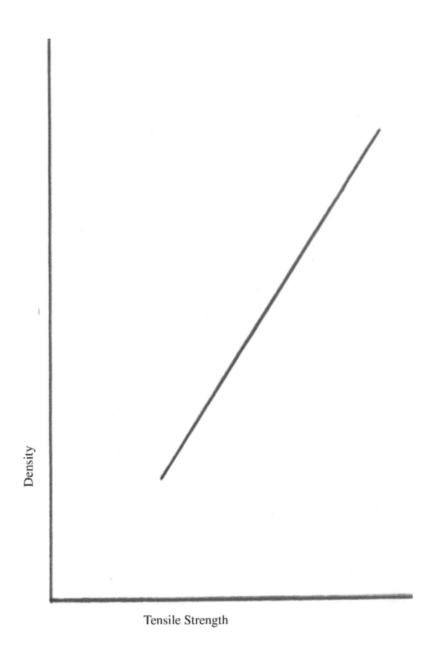

Figure 4-15 DENSITY VERSES TENSILE STRENGTH
The usual relationship between density and tensile strength for either a ceramic or a glass is the linear plot shown above. Many factors can cause this to deviate from linearity.

115

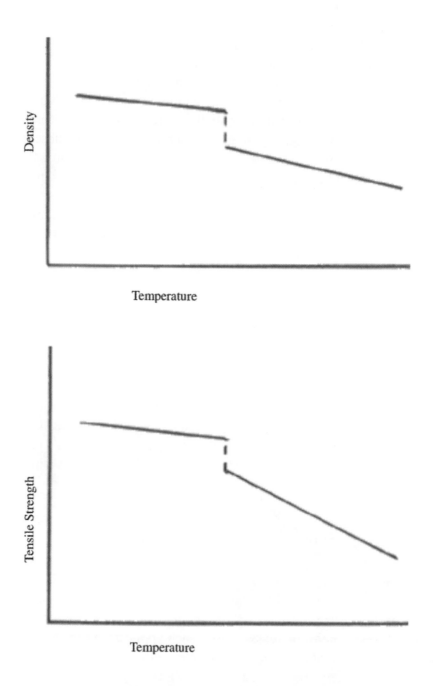

Figure 4-16 DENSITY AND TENSILE STRENGTH VERSES TEMPERATURE
If the linear relationship between density and tensile strength holds for a ceramic
than it usually has the above relationship to temperature. In both of the above graphs
a phase change is indicated by dotted lines. A similar plot, without the phase
change, would be expected for a glass.

116

THE EFFECT OF CRYSTAL STRUCTURE ON TENSILE PROPERTIES

The shorter the bond length the stronger the bond. Thus, a more closely packed ceramic will have a higher bond strength and a higher overall strength than the same ceramic in a less closely packed structure. Unless unpaired inner shell electrons are present, ceramics display their closest packing at low temperature with each subsequent phase change moving to an ever more loosely packed structure. Based solely on crystalline structure, the lower the temperature the stronger the ceramic.

Within a particular crystalline structure, the most closely packed direction will be the strongest direction. For multicrystalline ceramics with a random arrangement of crystal alignment, the measured strength is an averaging of the strengths in the direction of the applied force. Some of the crystal alignments will be such that the strength will be above average and these will not be beyond the breaking point of the bonds involved. However, some of the crystalline grains will be in weaker directions and these may be beyond their bond strength. This condition supports microcracking.

Aligned ceramics such as pyrolytic graphite and single crystals have a specific crystalline direction relative to the applied force. If this direction is strong than the measured tensile strength is well above the usual multi-grained ceramic of that type. The advantages of alignment are of prime design consideration when single crystal whiskers are used to strengthen a composite.

THE EFFECT OF CRYSTALLINE DEFECT STRUCTURE ON TENSILE PROPERTIES

A perfect crystal with no defects or grain bounties would be the strongest possible arrangement for any given chemical composition. Aside from carefully prepared ceramics near absolute zero, the ideal is never approached. With an increase in temperature, the amount of vacancies increases. With each application of force, dislocations are generated. With the slightest deviation during forming, grain boundaries are produced.

All of the various forms of crystal defects detract from the strength of chemical bonding within the ceramic. This would result in a very weak material if there were not some compensating aspects. Added vacancies allow mobility such that cracks can be blunted. Dislocations retard or prevent slipping within the grain. Grain boundaries mean a multi-grained material with an overall higher strength than the weakest direction within a single crystal.

How much benefit may be derived from what defect relative to a specific ceramic is very difficult to either control or predict. Generally a ceramic with less defect structure will have a higher tensile strength if the grain size is sufficiently small to support a random array relative to the applied force. If the grain size is large, the ceramic may suffer from grain "fall-out". This happens when a grain is so large its grain boundary serves as a breaking point in the cross-section. Figure 4-17 illustrates this.

THE EFFECT OF DEFECTS IN GLASS ON TENSILE PROPERTIES

The greatest source of defects leading to failure occur on the surface of the glass. Cracks, partially bonded atoms and weakly bonded atoms are more likely to cause further damage on the surface than if the same conditions occurred within the bulk of the glass. In the bulk the surrounding glass helps to compensate for these flaws. In addition, the bulk provides many paths for the defect to propagate. On the surface flaws are unrestrained on one side and only able to crack into the bulk. Thus surface flaws produce longer and more directed cracking than the same flaw within the bulk of the glass. See Figure 4-18.

Any process that can reduce the surface defect structure of a glass will greatly increase its tensile strength. The most utilized means is fire-polishing. Fire-polishing can literally be done with a flame but any heat source is effective. The surface of the glass is heated to above the softening point to allow flow. Rough spots on the surface will be eliminated and the added heat may encourage further bonding. If surface crystals are present, the glass can be further strengthened by heating the surface to above the melting point of the crystals. Either way a quick cooling is necessary to avoid both crystallization and corrosion.

Chemical etching, the deliberate corrosion of the surface of a glass, allows for smoothing of the surface because the more exposed areas will be the first to be dissolved. Chemical etching removes material, distorts the optical clarity of the glass, and tends to leave undesirable residuals. However, it can be done while the glass is in service, which is not usually possible with fire-polishing.

Additional strengthening tactics require altering the basic com-

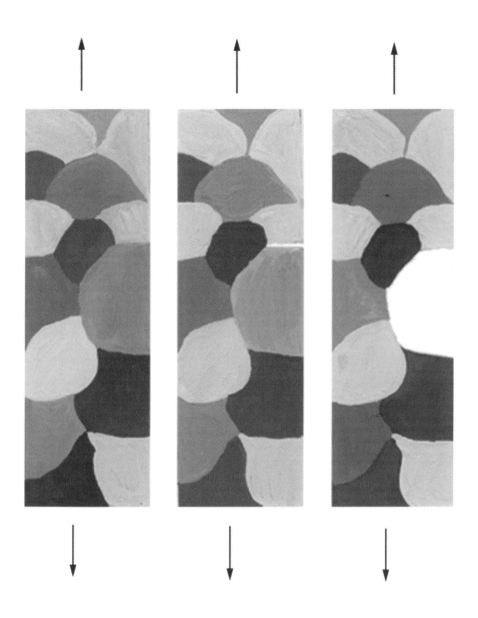

Figure 4-17 PULL OUT OF A GRAIN IN TENSION
The figure on the left depicts a ceramic at the moment a tensile load is applied. The
center figure depicts the same ceramic some moments later when the weakest of the
grain boundary bonds has broken. This break, which has been exaggerated for vis-
ibility, is shown on the right side just above center. Once a grain boundary begins
to separate the remaining bonds are severely strained and the whole grain can break
free of the bulk. This is shown in the figure on the right. Because there are few
grains in the cross section of this ceramic, the loss will probably cause it to break in
the vicinity of the lost grain.

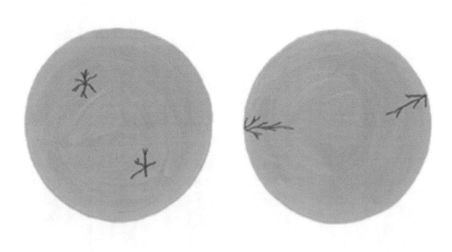

Figure 4-18 CRACKING IN GLASS DUE TO DEFECTS
These figures depict glass loaded perpendicular to the cross-sections shown. The
figure on the left has two defects causing cracking, which radiate out in a fairly uniform way. The figure on the right has the same crack producing conditions. Here
the cracks are longer both because they are limited in direction and the energy needed to advance the cracks reduces as their length increases.

position or structure of the glass. Processing can put the surface of the glass in compression. A compressive coating can also be obtained by atom substitution. Although a metallic coating will not always put the surface of a glass in compression, it will protect the glass, smooth the surface and provide additional bonding.

MECHANICAL PROPERTIES

CREEP IN CERAMICS

Creep is produced in a ceramic by one or more of three mechanisms: diffusion, dislocation motion, and grain boundary sliding. Diffusion is an especially effective mechanism for fine-grained ceramics at low stress and high temperature. The large volume of looser bonding along the grain bounties allows for greater mobility than could be obtained through the bulk of the grain. Both dislocation motion and grain boundary sliding are operative only in very specific materials.

At very low temperatures creep is not a consideration in ceramics. This temperature range will not support mass transport without the usual accompanying cracking. Thus, the ceramic does not creep; it slowly increases its defect structure until it crumbles.

When the temperature is sufficient to support diffusion, the ceramic is able to use the flow of vacancies to blunt cracks and prevent immediate failure in response to force. In addition, mass transport can be supported allowing for rearrangement relative to the force. As the free space within the ceramic increases, the ability of the ceramic to rearrange increases. The ceramic is then able to withstand either a larger force or the same force for a longer duration. While the resistance to failure is improved at a higher temperature, the creep rate is unfortunately increased.

Grain boundaries serve as better paths for diffusion than free space within the grains. As the grain size goes down the amount of pathways increases, which increases the diffusion rate. However, grain boundaries are a favored site for impurities. Impurities block diffusion and negate the effect of the grain boundaries. In practice the size of the grain is not usually significant for creep rate in compression. Small grain size promotes creep in tension. Large grain size is detrimental to tensile creep due to the tendency for grain fall-out leading to failure.

Ceramics with a large surface area such as fibers have the easiest diffusion path. They are also likely to have a more aligned structure than the usual random distribution relative to the force. High diffusion and internal slip combine within a fiber to produce a higher creep rate, which is compensated by a higher strength at failure than the parent ceramic.

Free space is not always an advantage to the creep strength of a ceramic. Increasing free space is often coupled with decreasing the amount and extent of bonding which degrades the internal strength. Highly porous ceramics are not as resistant to either creep or failure as would be expected when the mass of the material is considered. The large amount of free surface area within the ceramic gives rise to complications due to corrosion except under very controlled conditions.

Increasing the temperature of a ceramic decreases the strength of the overall bonding resulting in a large increase in creep. Creep is promoted both by a decrease in internal strength and by the increase in diffusion due to more vacancies. The elongation before breaking is especially noticeable if some ductility is possible in the ceramic.

Increasing the force in a temperature range where creep is operable results in a moderate to large increase in creep depending on how close the force is to the elastic limit. As the force approaches the elastic limit the ceramic behaves as it would in a tensile test, or a compression test depending on the mode of loading. Under these conditions, it will have a short time to failure and a moderate amount of creep because the time for mobility is limited. Lowering the force increases the mobility, which extends the time for creep but lowers the impetus for diffusion. A maximum creep for a specific temperature will be found at some mid range of force. This maximum point will most likely not be the same for compression and tension.

Increasing the surface roughness, impurity level, porosity or microcracking will usually increase creep about as much as increasing the force when the ceramic is in compression. The effect of any of these on tensile creep is such that failure occurs before diffusion can be significant except at a temperature approaching the melting point of the ceramic. However, there are many exceptions to this. For example, impurities may actually reduce creep if they have the effect of strengthening the ceramic.

Irradiation produces vacancies within the ceramic, which promote diffusion and creep in both tension and compression. Irradiation can also produce voluminous daughter products that stress and strengthen the ceramic, thus reducing creep and increasing failure strength in compression. These daughter products are usually not bound to the ceramic and therefore reduce both the failure strength and the creep in tension.

High temperature creep in simple ceramics, such as metal oxides, is a combination of plastic deformation due to dislocation motion within the grains aided by grain boundary sliding. Only when the bonds between the metal and the non-metal are sufficiently weak that they can be broken and reformed with the available energy will the ceramic be able to creep in this mode and retain its strength. It effectually becomes the metal component in its behavior. As such, it will be approximately as resistant to creep in both the tensile and the compressive mode. If the bonds are stronger, the ceramic begins to suffer localized internal cracking at the meeting point of multiple grain boundaries. Some of this cracking is blunted if not repaired by local diffusion. In this case the ceramic will be more resistant to creep in compression than in tension. This is due to the difference in crack propagation of these two loading modes.

Plate structures like clay and graphite deform in ways that resemble true creep but are actually more like a transformation in that they change the structural organization within the material. The mechanism is considered grain boundary sliding. If this process is carried out under conditions that allow the distortions to be minimized by diffusion than the deformation may be accomplished with minimal damage to the ceramic.

CREEP IN GLASS

Glass is subject to creep under its own weight at all but very low temperature. As the temperature increases the creep will increase despite the lack of appreciable diffusion. Creep is sustained because the glass can deform slightly in other ways. This is especially true of glass with limited bridging. As the temperature is further raised, diffusion becomes an increasingly important means of mass transport. Glass with a high porosity is especially prone to creep due to diffussional rearrangement.

In general, creep proceeds at a faster rate in a glass in tension than in compression. This is primarily due to the opening of the structure as it is stretched in tension. It is not a true plastic flow and the volume is not conserved. When the glass is in compression, voids are constricted and this inhibits movement as well as compacting the structure.

CREEP IN GLASS-CERAMICS

In glass-ceramics it is the glass that will most influence the rate

125

of creep. The ceramic will tend to retain its original shape while the glass flows around it, even if the adhesion between them is good. This leads to voids forming at the interface of the two materials. These voids are partially filled due to diffusion, if the temperature is high enough to support it. Voids do reduce the strength and increase the creep but they do not usually induce a failure due to the blunting nature of the diffused glass. See Figures 5-1 and 5-2.

If the glass-ceramic is in tension sufficient to overcome the compressive layer of the glass, it will fail at a shorter time and at a lower stress than would be expected. If the compressive layer is not compromised, the overall creep performance of the glass-ceramic will usually be a middle value between that of the glass and the ceramic.

In compression, the ceramic inhibits the flow of the glass and reduces the creep but it may also cause the material to shatter sooner. The two materials will not have the same response to the force and this imbalance will create stress concentrations at the interface. Usually the stress will cause localized cracking despite the compressive layer in the glass.

HIGH TEMPERATURE FATIGUE

Of the possible loading types in fatigue, the most common for ceramics and glasses is the compressive load followed by unloading. Both ceramics and glasses are the strongest in this loading type. Tensile loading followed by unloading is the least desirable type of fatigue loading given the propensity for crack propagation in tension. A combination of tensile and compression will yield a fatigue life between the other two loading types.

Ceramics and glasses respond to cyclic loading at high temperature in a similar way to that of creep. The mechanisms are the same, especially if the load is maintained for a duration of time. The time span without a load usually prolongs the time to failure as it allows diffusion to blunt cracks and relieve residual stresses. However, the added time without a load may also allow additional corrosion. Compression slows diffusion of the corroding material.

While it would seem that fatigue life of a combination tensile-compressive cycle would be mid value between the compressive fatigue life and the tensile fatigue life, the actual life is generally longer. This is due to the positive effects of the compressive cycle. Compression allows ceramics to relieve stress without encouraging internal corrosion or crack propagation.

Conditions that produce cyclic loading are often accompanied by vibration. Ceramics are especially sensitive to vibration; glasses are less responsive. Because of their high chemical bonding, ceramics transmit most wavelengths of vibration easily. Certain frequencies are absorbed and stored as energy. This additional energy can be used in positive ways such as reduction of cracking or it can be used in negative ways such as promotion of cracking. What energy will be absorbed and how it will be used is not usually predictable from bonding and structure.

LOW TEMPERATURE FATIGUE

Whereas creep data is useful in predicting both fatigue life and mechanisms at high temperature, it is either not available or not useful at temperatures below half of the melting point of a ceramic. Failure mechanisms in this range are generally assumed to be those of the tensile type. Ceramics are inspected for cracking in a variety of ways but none have proven successful in predicting failure. Glass-ceramics share this problem.

Creep data for glass does exist at low temperature because it is an on going problem. Glasses more readily reveal cracks that will culminate in failure. Visual inspection is reasonably effective in all but opaque glasses. Too much reliance should not be placed on visual inspection of glass as it ignores the significant speed at which cracking can sometimes occur. A combination of creep analysis and visual inspection is prudent.

WEAR IN CERAMICS

Most of what is assumed to be wear in ceramics is actually a crushing of the surface which flakes away. This is undesirable in the sense of loosing material. However, the flaking reduces the risk of build up on the wear surface if the ceramic is not in a confined space. Wear resistance is directly related to the strength of the ceramic. Because wear is often a combination of forces, both the compressive and the tensile aspects of strength should be considered.

The movement of the wear surface imparts added residual stress, which aids corrosion. In addition, wear causes uneven surface removal which encourages grain pull-out. Grains with less dense packing orientated along the surface of the ceramic will suffer the most degradation.

Also of concern is the nature of the wear-inducing surface. In

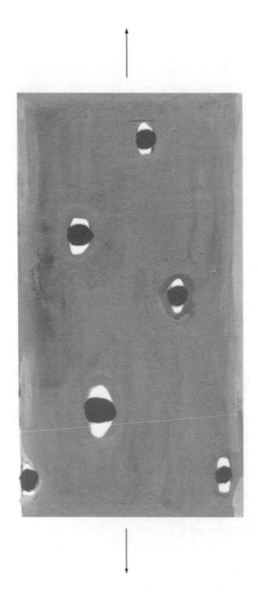

Figure 5-1 VOID FORMATION DURING TENSILE CREEP
The glass in this figure is the light gray bulk whereas the ceramic is the darker gray. The white areas above and below the ceramic are the voids. These voids are probably bigger than the expected voids produced within a glass/ceramic composite of this type as most desirable composites will display better bonding.

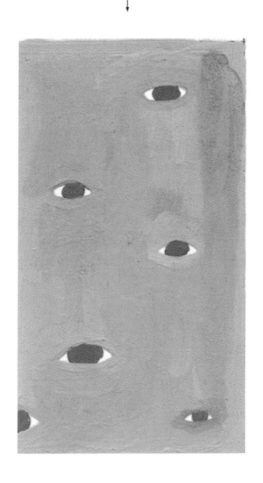

Figure 5-2 VOID FORMATION DURING COMPRESSIVE CREEP
As in Figure 5-1, this is a depiction of a glass-ceramic. In most glass-ceramics the ceramic would be both more voluminous and more irregularly shaped. The voids are exaggerated for ease of identification. In this orientation the voids are less of concern because the compressive force tends to constrict them.

129

general a strong, rough, dry sliding piece with impact loading and a twist to the movement will do far more damage than a weak, smooth, lubricated sliding piece with an even movement. Ceramics, being the complex materials they are, do not always behave as expected. The lubricant may cause swelling and cracking. The smooth surface may grind the flakes into the ceramic more than the rougher surface, which will collect the flakes. The only thing that is reasonably certain is that impact loading is detrimental to ceramics.

WEAR IN GLASS

Glass is a more homogeneous material than ceramic and not subject to localized degradation such as grain pull-out. Glass is also not subject to flaking which can be so detrimental to the wear resistance of a ceramic. The major danger with glass is that the wear inducing surface will penetrate the protective compressive layer. If this happens to even a small area of the glass, the glass may quickly shatter.

An additional problem for glass, which is usually not encountered with a ceramic, is sufficient heating to cause localized flow of the surface. Generally the materials are of high enough melting temperature that they do not actually melt, but glass can flow and distort the working surface. The heat build up may also detemper the glass leaving it venerable.

WEAR IN GLASS-CERAMICS

The combination of a glass and a ceramic can often prove to be excellent in wear resistance. The smooth, homogeneous glass surface encourages easy sliding. The strengthening of the ceramic provides added support and inhibits flow of the glass. The glass protects the ceramic form grain pull-out and other forms of uneven wear as well as corrosion. The ceramic sufficiently reinforces the glass such that the compressive layer is unlikely to be breached. If it is, the presence of the ceramic will prevent immediate shattering.

IMPACT STRENGTH

Impact strength is difficult to predict. Glass is usually better behaved than ceramic but both have surprises in their nature. Glass-ceramic is often superior to either glass or ceramic in impact strength. This is especially true if the glass-ceramic is further reinforced with fibers to add to its fracture toughness.

Ceramics can often have high impact strength if they are

designed for it. Both high strength and high fracture toughness are desirable in an impact resistant ceramic, which leads to compromises in microstructure. A moderate grain size will be more favorable than the usual fine grain size used to increase strength. Weaker grains such as those with high porosity are desirable because that also encourages fracture toughness at the expense of strength.

Glass has a low modulus of elasticity, which makes it more resilient to impact. However, it also has a low fracture toughness so damage caused by impact is easily spread. This spread can take on some strange patterns. For instance, a plate glass window struck in the center can have the corners break.

FRACTURE TOUGHNESS

What ever prevents cracking or slows its progress will aid the fracture toughness of a material. Often this is at the expense of other desirable characteristics such as strength, clarity or wear resistance. Compromises in performance for the sake of longevity are to be expected with any material but this is particularly true of ceramics.

Preventing a crack from forming in a ceramic is generally not possible except for the most carefully prepared whiskers. Anything that will blunt a crack such as void formation, diffusion, and corrosion of the tip will often stop a crack from propagation.

When blunting is not effective, the dissipation of the stress driving the crack is the next line of defense. If the ceramic is in tension, an unloading will help at this stage. However, the larger the crack the less energy it requires to continue, and very large cracks can become self-propagating.

Heating to above half of the melting point to relieve residual stress not only removes energy that could be used to propagate a crack, it encourages those things that produce blunting. Whenever possible, this is an effective life extension procedure if corrosion can be controlled.

Large grain size requires the crack to either take a non-direct route along grain boundaries or go through a grain. Both require more energy then a direct path through a finely grained ceramic. Cracking within the grain encourages branching of the crack, which also consumes energy. Branching comes about because the crack will follow crystallographic lines. The least densely packed plane is the most favorable but it is generally not the plane aligned with the advancing main crack. See Figure 5-3.

Inclusions, impurities and fillers can all serve as barriers to an advancing crack. However, strength enhancing materials such as whiskers are generally more desirable and more effective. Glass-ceramics can be effective in this due to the blunting effect of the glass component throughout the material.

Glass is less sensitive to cracking than ceramics. Preventing a crack from forming in a glass is done by tempering. If a glass is not tempered or has lost it tempering during usage then it is generally too fragile to be of service.

Fracture toughness is also sensitive to the shape and size of the part as well as the surface condition and mode of use. Making a part out of a candidate material and trying it in service is the only certain means of evaluation. Otherwise consider what is known of the impact strength, tensile strength and hardness, and pray for divine guidance.

HARDNESS

Microhardness measurements show a wider variation within a ceramic than do macrohardness measurements. This difference is due to the "averaging" nature of the macrohardness measurement. Usually macrohardness is used for general sorting of minerals whereas microhardness is a research and development tool.

For microhardness the testing point is sufficiently small that the entire test is conducted on one grain or one segment of grain boundary. Thus microhardness values are sensitive to variations in crystallographic planes. Within a plane the packing factor will reflect the hardness, as the hardness is almost wholly dependent on the strength of the chemical bonds holding the ceramic together. Fewer bonds and longer bond lengths mean lower bond strength and lower hardness. Similarly in closed packed planes the bond numbers and directions are favorable to high strength and high hardness. The hardness declines approximately linearly with increasing porosity and deviation from stoichiometry. Hardness declines with temperature as Figure 5-3 illustrates.

In metals, the macrohardness test is often used as a quick indicator of the tensile strength. It is also considered a non-destructive test. Neither of these assumptions can be made with respect to ceramics. While the tensile strength of a ceramic does depend on the bond strength, it is also highly influenced by other factors that only slightly modify macrohardness values. The two most notable of these are porosity and deviation from stoichiometry. In addition, the use of an

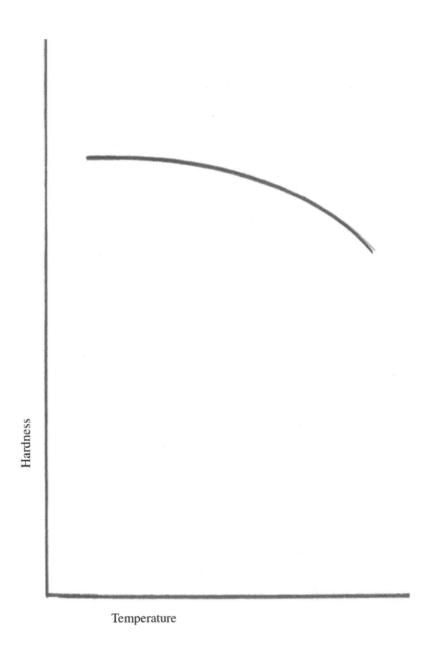

Figure 5-3 HARDNESS AS A FUNCTION OF TEMPERATURE
The gradual slump in hardness as a function of temperature is the usual response for ceramics, glasses and glass-ceramics. However, many factors can change this behavior.

indentor crushes the brittle surface of the ceramic producing cracks that can lead to premature failure of the material. Only if a ductile transition occurs may the ceramic be considered like a metal with respect to macrohardness assumptions.

DATA

ALUMINUM FERRATE, EMERY, $Al_2O_3 \cdot Fe_3O_4$

SPECIFIC GRAVITY: 3.7-4.3
HARDNESS: 1400 Knoop; 8 Mohs
WEAR RATE: 0.055-0.075 mm.mm/kg.km on steel
COLOR: dark red to brown
TOXICITY: Only the dust form is an irritant.
STRUCTURE: Emery is basically impure alumina and displays the characteristics of alumina for most properties. Interspersed within the grains of alumina is iron oxide which is not intimately chemically joined to the alumina.
APPLICATIONS: Originally emery was used extensively as an abrasive especially in grinding wheels. Now the greatest volume is used in non-skid stair treads.

ALUMINUM NITRIDE, AlN

MAXIMUM CONTINUOUS SERVICE TEMPERATURE: 1100° C in air; 1800° C in an inert atmosphere

DECOMPOSITION TEMPERATURE: about 2000° C
SPECIFIC GRAVITY: 3.26

SPECIFIC HEAT: 1.7 kJ/kgK for 0° to 100° C, 2.0 Kj/kgK for 100° to 600° C

THERMAL CONDUCTIVITY: 185 W/mK at 20° C, 29.2 W/mK at 200° C, 2.1 W/mK at 800° C; superconductive below 1.55 K

COEFFICIENT OF LINEAR EXPANSION: 5.6×10^{-6}/K over 20° to 1000° C
THERMAL SHOCK: moderate resistance
HARDNESS: 1225 Knoop; 6.5 Mohs
MODULUS OF ELASTICITY: 345,000 MPa at 20° C, 317,400 MPa at 1000° C, 276,000 MPa at 1400° C; 350,000 MPa for a whisker
TENSILE STRENGTH: 6,900 MPa for a whisker
RUPTURE MODULUS: 265 MPa at 20° C, 136 MPa at 1000° C, 125 MPa at 1400° C

FLEXURAL STRENGTH: 262 MPa at 20° C, 172 MPa. at 1095° C
FRACTURE TOUGHNESS: 2.6 MPa.m$^{1/2}$
COMPRESSIVE STRENGTH: 20,700 MPa
WEIBULL MODULUS: 20
POISSON'S RATIO: 0.25
ELECTRICAL RESISTIVITY: It is superconductive below 1.55 K. 10^{11} Ohm.cm at 20° C
DIELECTRIC STRENGTH: 10 kv/mm

DIELECTRIC CONSTANT: 8.9 at 1 MHz, 8.2 at 7 GHz
DIELECTRIC LOSS: 0.001 at 1 MHz, 0.002 at 7 GHz
BAND GAP: 6.2 eV
ABSORPTION EDGE: 0.12 microns
REFRACTIVE INDEX: 3.32
COLOR: white
SOLUBILITY IN WATER: slight
CHEMICAL STABILITY; RESISTANCE TO:

WATER: Aluminum nitrate will react with water or steam to form toxic, corrosive fumes.

GASES: Oxidation usually occurs around 1000° C.

METALS: It is stable in contact with aluminum to 2000° C and most other metals up to 1200° C.

TOXICITY: It is not toxic in itself but it easily forms ammonia.

STRUCTURE: hexagonal

APPLICATIONS: Aluminum nitride is used where light weight is desirable, especially in areas where moisture or hydrogen are not present.

ALUMINUM OXIDE, ALUMINA, CORUNDUM, RUBY, SAPPHIRE, alpha phase
of Al_2O_3

MAXIMUM CONTINUOUS SERVICE TEMPERATURE: 1650° C; 1250° C for fiber form

MELTING TEMPERATURE: 2050° C

BOILING POINT: 2980° C

MINIMUM MELTING POINT OF BINARY SYSTEM: 1835° C with beryllium oxide; 1360° C with calcium oxide; 1995° C with magnesium oxide; 1546° C with silica; 1750° C with thorium oxide; 1720° C with titania; 1710° C with zirconia; 1310° C with ferrous oxide; 1425° C with barium oxide; 2030° C with chromium oxide; 1520° C with manganese oxide; 1880° C with nickel oxide; 865° C with lead oxide; 1505° C with strontium oxide; 1710° C with zinc oxide
SPECIFIC GRAVITY: 4.0-4.1 for alpha form, 3.5-3.9 for gamma form
SPECIFIC HEAT: 0.54 kJ/kgK at 73 K, 0.21 kJ/kgK at 123 K, 0.40 kJ/kgK at 173 K, 0.58 kJ/kgK at 223 K, 0.82 kJ/kgK at 20° C, 1.25 kJ/kgK for 1100° to 2300° C
THERMAL CONDUCTIVITY: 126 W/mK at 10 K, 587 W/mK at 40 K, 36 W/mK at 20° C, 8 W/mK at 100° C, 12 W/mK at 400° C, 7 W/mK at 800° C, 1 W/mK at 1000° C, 5 W/mK at 1200° to 1600° C, 1 W/mK at 1900° C for alumina; 27 W/mK for fiber form; 69 W/mK at 20° C, 7 W/mK at 1100° C for sapphire
COEFFICIENT OF LINEAR EXPANSION: 9.5×10^{-6}/K from 0° to 800° C
THERMAL SHOCK: moderate resistance

HARDNESS: 2100 Knoop or 9 Mohs at 20° C, 1300 Knoop at 540° C, 230 Knoop at 1220° C, 80 Knoop at 1650° C for alumina

ELONGATION: 0.4% for fiber of alumina

MODULUS OF ELASTICITY: 380,000 MPa for fiber; 290,000-370,000 MPa for ruby and sapphire from 25° to 1200° C; 370,000-407,000 MPa for alumina from 25° to 1200° C; 511,000 MPa for sapphire whisker.

MODULUS OF RIGIDITY: 163,000 MPa for alumina; 166,000 MPa for sapphire

TENSILE STRENGTH: 223 MPa at 1100° C, 90 MPa at 1200° C for alumina; 1,400-2,100 MPa at 20° C, 1,100 MPa at 1100° C for fiber; 173 MPa at 20° C, 345 MPa at 600° C, 504 MPa at 1000° C, 635 MPa at 1200° C for sapphire; 117,000 MPa for sapphire whisker RUPTURE MODULUS: 345-704 MPa at 25° C, 197-324 MPa at 600° C and 310-590 MPa at 1000° C for sapphire and ruby. Most of the range comes about due to variation in orientation. 242-462 MPa at 25° C, 235-359 MPa at 400° C, 214-338 MPa at 1000° C, and 97-255 MPa at 1350° C for alumina. Most of the variation comes about due to grain size.

IMPACT STRENGTH: 0.8 J over 0° to 1000° C for alumina

FLEXURAL STRENGTH: 400 MPa at 20° C, 255 MPa at 800° C, 242 MPa at 1200° C for alumina; 518-760 MPa for sapphire

FRACTURE TOUGHNESS: 4 MPa.m$^{1/2}$

FRACTURE ENERGY: 30 J/m.m

SHEAR STRENGTH: 345 MPa

SHEAR MODULUS: 164,000 MPa

COMPRESSIVE STRENGTH: 260-450 MPa for alumina; 3,060-3,420 MPa for sapphire; 2,070 MPa for ruby; 3,000 MPa at 25° C, 1,500 MPa at 400° C, 1,300 MPa at 800° C, 900 MPa at 1000° C, 500 MPa at 1200° C, 250 MPa at 1400° C and 50 MPa at 1600° C for very pure alumina

CREEP RATE: 1.3×10^{-6}/hr at 1300° C, 7.5×10^{-4}/hr at 1360° C for alumina with a 13 MPa load; 7.2×10^{-6}/hr at 1300° C for sapphire with a load of 100 MPa

WEIBULL MODULUS: 10

WEIBULL NUMBER: 450 MPa

FAILURE STRESS RANGE: 300-500 MPa

FATIGUE: rolling-contact life for alumina balls at 10^6 cycles to failure, speed 950 rpm, contact angle 20°, temperature 27° C, and a lubricant of mineral oil. The results are:

Material	Hertzian Stress, MPa	50% failure	90% failure
hot-pressed	3.5	5	8
hot-pressed	3.8	1.3	2.1
hot-pressed	4.1	1.3	2.8
hot-pressed	4.5	0.3	0.4
cold-pressed	1.7	7.5	17
cold-pressed	2.1	1.6	2.0
cold-pressed	2.4	0.5	0.5

POISSON'S RATIO: 0.26-0.32 at 25° C, 0.32 at 1000° C and 0.45 at 1400° C for alumina

COEFFICIENT OF FRICTION: 0.95 against nickel; 0.73 against nickel plus water for alumina; 0.2 for sapphire on sapphire, 0.15 for sapphire on steel

WELDING ON TEMPERATURE: over 1600° C on steel

WEAR RATE: 0.002-0.01 mm.mm/kg.km on steel; 0.0017-0.0039 mm.mm/kg.km on gray cast iron for alumina

ELECTRICAL RESISTIVITY: 10^{16} Ohm.cm from 20° to 200° C, 10^{10} Ohm.cm at 500° C, 10^6 Ohm.cm at 1000° C for alumina; 10^{11} Ohm.cm at 500° C, 10^6 Ohm.cm at 1000° C, 10^4 Ohm.cm at 1500° C and 10^3 Ohm.cm at 2000° C for sapphire

DIELECTRIC STRENGTH: 10 kv/mm

DIELECTRIC CONSTANT: 7 at 100 to 10^{10} Hz for alumina; 10.5 at 10^5 Hz for fiber

DISSIPATION FACTOR: 0.002 at 100 to 10^{10} Hz for alumina; 0.0001 at 100 to 10^{10} Hz for sapphire

MAGNETIC SUSCEPTIBILITY: -4.40×10^{-7} m^3/kg for alumina; natural rubies are diamagnetic but synthetic rubies are paramagnetic.

REFRACTIVE INDEX: 1.76-1.77

BIREFRINGENCE: 0.008

TRANSMISSION OF LIGHT: at least 10% of incident light is transmitted through a 2 mm. thickness in the region of 0.2-7 microns for alumina and 0.15-7.5 microns for sapphire

COLOR: white for alumina, red for ruby and blue for sapphire

ACOUSTIC LONGITUDINAL WAVE VELOCITY: 10.52 km/s for alumina, 11.1 km/s for sapphire

ACOUSTIC SHEAR WAVE VELOCITY: 6.04 km/s for sapphire

ACOUSTIC LONGITUDINAL WAVE IMPEDANCE: 40.6 kg/m^{2}s for alumina, 44.3 kg/m^{2}s for sapphire

WATER SOLUBILITY: 10^{-4} g/100ml at 29° C

WATER ABSORPTION: 0% for fully dense material; 5-15% for 2.4 specific gravity

GAS TRANSMISSION RATE: less than 4×10^{-13} g/c-m.sec for sapphire from 20° to 2015° C

DIFFUSION COEFFICIENT: 10^{-14} cm.cm/sec at 1400° C and 10^{-12} ctn.cm/sec at 1650° C for oxygen; 4×10^{-12} cm.cm/sec at 1500° C and 2×10^{-10} cm.cm/sec at 1800° C

CHEMICAL STABILITY; RESISTANCE TO:

    ACIDS: Pure alumina is extremely resistant to acids except to hydrofluoric acid.

    BASES: It deteriorates only after long exposure.

    GASES: There is a slight reaction with hydrogen fluoride and hydrogen sulphide; otherwise it is very inert.

METALS: Liquid titanium attacks alumina above 1725^0 C.

TOXICITY: It is nontoxic unless inhaled in the form of very fine powder.

RADIOACTIVE EMISSIVITY: 0.11 at 200^0 C, 0.15 at 400^0 C, 0.19 at 600^0 C

STRUCTURE: Alumina is hexagonal close packed in gamma phase and trigonal in alpha or corundum phase. The gamma phase transforms to the alpha phase at 1150-1200^0 C. This phase change is not reversible on cooling.

ADDITIVES: Nickel oxide, zinc oxide, cobalt oxide or tin oxide can all be added to alumina up to about 0.25% with the same result. These oxides are insoluble in alumina so they all aid densification during sintering without entering into the matrix of the alumina. Magnesium oxide is added up to about 0.5% to suppress discontinuous grain growth. The addition of up to 10% titania assists in the sintering process resulting in a uniform fine grained microstructure of near theoretical density. The normal gem form of alumina is sapphire but when about 0.75% chromium oxide is present it becomes a ruby.

APPLICATIONS: Since alumina is a strong, hard, cheap oxide its uses are extensive and continue to expand. Some current applications are as a refractory in molten metal containment, in insulating applications such as spark plugs, as tool blades and as an abrasive in polishing powders. The high purity grade finds medical applications in surgical implants. The ruby form has been used in lasers and sapphires have also found similar applications. Sapphire is also used as a substrate in microelectronic circuits and for specialty optical flats and windows. Ruby and sapphire are considerably more expensive than alumina.

ALUMINUM SILICATE, ANDALUSITE, CYANITE, SILLIMANITE, MULLITE, $Al_2O_3 \cdot SiO_2$

MAXIMUM CONTINUOUS SERVICE TEMPERATURE: 1100^0 C

MELTING TEMPERATURE: 1920^0 C

SPECIFIC GRAVITY: 3.16 for mullite; 3.2 for sillimanite, andalusite and cyanite

SPECIFIC HEAT: 1.0 kJ/kgK

THERMAL CONDUCTIVITY: 0.2 W/mK

COEFFICIENT OF LINEAR EXPANSION: $5.3X10^{-6}$/K for 20^0 to 1100^0 C

THERMAL SHOCK: good resistance

HARDNESS: 7.5 Mohs for andalusite and sillimanite

MODULUS OF ELASTICITY: 220,000 MPa at 20^0 C, 48,000 MPa at 1200^0 C

TENSILE STRENGTH: 84 MPa

RUPTURE STRENGTH: 185 MPa

FLEXURAL STRENGTH: 132 MPa This remains constant unt5il about 1400^0 C. Values as high as 350 MPa have been reported for very selective sintering conditions.

SHEAR STRENGTH: 131 MPa

COMPRESSIVE STRENGTH: 676 MPa

FRACTURE TOUGHNESS: 2.2 MPa.m$^{1/2}$

CREEP RATE: 9.8×10^{-9}/hr at 1100° C and a load of 8 MPa
POISSON'S RATIO: 0.25
ELECTRICAL RESISTIVITY: 10^{16} Ohm.cm at 20° C, 10^{10} Ohm.cm at 500° C
DIELECTRIC STRENGTH: 10 kv/mm
DIELECTRIC CONSTANT: 4.1 at 1 MHz
DISSIPATION FACTOR: 0.003 at 1 MHz
REFRACTIVE INDEX: 1.66
BIREFRINGENCE: 0.021 for sillimanite; 0.016 for cyanite; 0.10 for mullite
COLOR: clear to cloudy
SOLUBILITY IN WATER: none
CHEMICAL STABILITY; RESISTANCE TO:
    ACIDS: good
    BASES: fair

    GASES: It reacts with hydrogen above 1100° C to form volatile products.
    METALS: Aluminum silicate has fair resistance to both solid and liquid form.
TOXICITY: Shipping regulations exist due to the silica content.
RADIOACTIVE EMITTANCE: 0.5
STRUCTURE: Andalusite, also known as alpha, is orthorhombic, cyanite is tri-
clinic and sillimanite is orthorhombic. They all transform to mullite at 1545° C.
Mullite is also orthorhombic but it has the chemical composition of $3Al_2O_3 \cdot 2SiO_2$
rather than the $Al_2O_3 \cdot SiO_2$ of the lower temperature forms.
APPLICATIONS: It is used as a glaze for alumina and as a refractory furnace lin-
ing.

ANTIMONY, Sb

MELTING TEMPERATURE: 630° C
BOILING TEMPERATURE: 1380° C
SPECIFIC GRAVITY: 6.62
SPECIFIC HEAT: 0.21 kJ/kgK
THERMAL CONDUCTIVITY: 18.9 w/mK
COEFFICIENT OF THERMAL EXPANSION: 8.5×10^{-6}/K
HARDNESS: 30 - 58 Brinell
TENSILE STRENGTH: 11 MPa
MODULUS OF ELASTICITY: 78,000 MPa
ELECTRICAL RESISTIVITY: 3.9×10^{-5} Ohm.cm
MAGNETIC SUSCEPTIBILITY: -10^{-8} m^3/kg
REFLECTIVITY: 70% at 0.6 microns
COLOR: silver
CHEMICAL STABILITY; RESISTANCE TO:
        BASES: Antimony is easily attacked by concentrated bases.
        ACIDS: It is attacked by hydrochloric acid, nitric acid and sulfuric acid
when they are concentrated.

GASES: Both air and chlorine attack antimony.
TOXICITY: Antimony is highly toxic. It also possesses a moderate radiation, fire and explosion hazard.
STRUCTURE: Cubic and rhombohedral low temperature forms transform into hexagonal.
APPLICATIONS: Antimony finds extensive use in storage batteries.

ANTIMONY TRIOXIDE, SENARMONTITE, VALENTINITE, Sb_2O_3

MELTING TEMPERATURE: 656° C

BOILING TEMPERATURE: 1550° C
SPECIFIC GRAVITY: 5.2 for senarmontite, 5.7 for valentinite
HARDNESS: 2.5 Mohs
MAGNETIC SUSCEPTIBILITY: $-8.72X10^{-7}$ m^3/kg
REFRACTIVE INDEX: 2.1 for senarmontite, 2.4 for valentinite
COLOR: clear to white
SOLUBILITY IN WATER: very slight
TOXICITY: It is an irritant on contact with skin or mucous membranes. It seems to have the same effects as lead or arsenic in causing digestive disturbance sometimes to the point of death.
STRUCTURE: senarmontite is cubic, valentinite is rhombic.
APPLICATIONS: It is used as an opacifier and coloring agent in glazes to produce white. With lead it yields Naples yellow and with tin the result is gray.

ARGON, A

MELTING TEMPERATURE: 84 K
BOILING TEMPERATURE: 87 K
SPECIFIC GRAVITY: 0.0018
SPECIFIC HEAT: 0.52 KJ/KgK
THERMAL CONDUCTIVITY: 0.0144 W/mK at 233 K, 0.0156 W/mK at 255 K, 0.0190 W/mK at 49° C
DIELECTRIC CONSTANT: 1.53 at 82 K
MAGNETIC SUSCEPTIBILITY: $-6.16X10^{-9}$ m^3/kg
COLOR: colorless gas
CHEMICAL STABILITY: inert
TOXICITY: Like the other inert gases, argon acts as an asphyxiant because it interferes with the assimilation of oxygen in the lungs.
STRUCTURE: face centered cubic
APPLICATIONS: Argon is extensively used when an inert atmosphere is desired, especially in laboratories.

ARSENIC, As

MELTING TEMPERATURE: 817° C at 28 atmospheres for the hexagonal form

DECOMPOSITION TEMPERATURE: 613° C for hexagonal form, 358° C for cubic form

SPECIFIC GRAVITY: 5.73 for hexagonal, 2.03 for cubic

SPECIFIC HEAT: 0.329 kJ/kgK

THERMAL CONDUCTIVITY: 50 W/mK

COEFFICIENT OF LINEAR THERMAL CONDUCTIVITY: $4.7X10^{-6}$/K

HARDNESS: 1440 Brinell

ELECTRICAL RESISTIVITY: $2.6X10^{-5}$ Ohm.cm

MAGNETIC SUSCEPTIBILITY: $-0.917X10^{-11}$ m^3/kg for rhombohedral, - $3.97X10^{-9}$ m^3/kg for hexagonal

COLOR: gray when hexagonal, yellow when cubic, silver when rhombohedral

REFRACTIVE INDEX: 1.00

SOLUBILITY IN WATER: none

TOXICITY: Arsenic is both highly toxic and a carcinogen.

FIRE HAZARD: It is a moderate hazard in dust form.

STRUCTURE: Arsenic is rhombohedral until 228° C when it transforms into hexagonal, cubic or amorphous.

APPLICATIONS: Arsenic is used in shot gun pellets and as a doping agent in transistors.

ARSENIC TRIOXIDE, ARSONOLITE, CLAUDETITE, As_2O_3

MELTING/DECOMPOSITION TEMPERATURE: 193° C

BOILING TEMPERATURE: 457° C for claudetite

SPECIFIC GRAVITY: 3.87 for arsonolite, 4.15 for claudetite

HARDNESS: 1.5 Mohs

REFRACTIVE INDEX: 1.76 for arsonolite, 1.87-2.01 for claudetite

COLOR: clear to white

SOLUBILITY IN WATER: 1.2 g/100ml in cold water and 11.46 g/100ml in hot water. Combining it with water may free the arsenic which is highly toxic.

TOXICITY: Arsenic trioxide is odorless and tasteless. Ingestion can cause immediate death. Exposure to it via inhalation or skin contact causes a wide variety of symptoms including contact area cancerous growth. The exposure effects are cumulative so that death is the final outcome.

STRUCTURE: arsonolite is cubic and claudetite is monoclinic

APPLICATIONS: It serves as a good opacifying agent in glazes.

ARSENIC TRISULFIDE, ORPIMENT, As_2S_3

MELTING TEMPERATURE: 300° C

BOILING TEMPERATURE: 707^0 C
SPECIFIC GRAVITY: 3.4
POISSON'S RATIO: 0.29
INDEX OF REFRACTION: 2.4-3.0
ACOUSTIC LONGITUDINAL WAVE VELOCITY: 2.58 km/s
ACOUSTIC SHEAR WAVE VELOCITY: 1.4 km/s
ACOUSTIC LONGITUDINAL WAVE IMPEDANCE: 8.25 kg/m^{2}s
SOLUBILITY IN WATER: in cold water 0.00005 g/100ml, no solubility in hot water.
TOXICITY: Ingestion can cause immediate death. Exposure to it via inhalation or skin contact causes a wide variety of symptoms including contact area cancerous growth. The exposure effects are cumulative so that death is the final outcome.
STRUCTURE: monoclinic
APPLICATIONS: It forms a glass which is translucent to infrared light.

ASBESTOS, CHRYSOTILE [$3MgO \cdot 2SiO_2 \cdot 2H_2O$], AMOSITE [$Fe/MgO \cdot SiO_3$],

CROCIDOLITE [$NaFe(SiO_3)_2 \cdot FeSiO_3 \cdot H_2O$],

ANTHOPHYLLITE [$(Mg/Fe)_7 \cdot Si_8O_{22} \cdot (OH)_2$],

TREMOLITE [$Ca_2Mg_5Si_8O_{22} \cdot (OH)_2$],

ACTOLITE [$CaO \cdot 3(MgFe)O \cdot 4SiO_2$]

MAXIMUM CONTINUOUS SERVICE TEMPERATURE: 125^0 C for woven forms, 600^0 C for other forms
DECOMPOSITION TEMPERATURE: 450^0 C for chrysotile; 300^0 C for crocidolite
SPECIFIC GRAVITY: 2.55 for chrysotile; 2.85-3.1 for anthophyllite; 3.43 for amosite; 3.37 for crocidolite; 3.1 for tremolite
SPECIFIC HEAT: 1.1 kJ/kgK for chrysotile; 0.9 kJ/kgK for anthophyllite; 0.8 kJ/kgk for amosite and crocidolite
HARDNESS: 2.5-4 Mohs for chrysotile; 5.5-6.0 Mohs for anthophyllite and amosite; 4 Mohs for crocidolite
MODULUS OF ELASTICITY: 165,000 MPa for chrysotile and amosite; 158,000 MPa for anthophyllite; 190,000 MPa for crocidolite
TENSILE STRENGTH: 4,600 MPa at 20^0 to 200^0 C, 3,700 MPa at 350^0 C, 1,200 MPa at 450^0 C, 700 MPa at 700^0 C for crocidolite; 3,700 MPa at 20^0 to 400^0 C, 4,100 MPa at 550^0 C, 300 MPa at 650^0 C for chrysotile; 2,600 MPa at 20^0 to 250^0 C, 800 MPa at 400^0 C, 200 MPa at 600^0 C for amosite; 2,400 MPa for anthophyllite; 5,000 MPa for tremolite
COEFFICIENT OF FRICTION: 0.45 for woven, slightly less for other forms.
WEAR RATE: 9.2X10^{-6} mm.mm.mm/J at 100^0 C for woven, slightly less for other forms

143

ELECTRICAL RESISTIVITY: 3×10^3 to 1.5×10^5 Ohm.cm for chrysotile; 2.5×10^6 to 7.5×10^6 Ohm.cm for anthophyllite; up to 5×10^8 Ohm.cm for amosite; 2×10^5 to 5×10^5 Ohm.cm for crocidolite MAGNETIC SUSCEPTIBILITY: 6.7×10^{-8} m^3/kg for chrysolite; 1.8×10^{-7} m^3/kg for anthophyllite; 9.9×10^{-7} m^3/kg for amosite; 7.7×10^{-7} m^3/kg for crocidolite

REFRACTIVE INDEX: 1.61 for anthophyllite, 1.67 for amosite, 1.70 for crocidolite, 1.54 for chrysotile

BIREFRINGENCE: moderate for chrysotile and anthophyllite; strong for amosite; weak for crocidolite

COLOR: chrysotile, tremolite and anthophyllite are white or clear, amosite is gray to brown, crocidolite is blue

CHEMICAL STABILITY; RESISTANCE TO:

ACIDS: highly resistant in general except chrysolite which is not very resistant.

BASES: good resistance

SOLVENTS: good resistance to organic solvents

TOXICITY: Small particles of asbestos dust can be inhaled. When this happens the dust is lodged in the lungs and fibrosis or scar tissue is formed around it. This part of the lung will not regenerate and the irritation may in time cause cancer. Anyone installing or otherwise cutting asbestos is especially at risk but just having it in building materials can be hazardous.

WEAKNESSES: crocidolite fuses at high temperature

STRUCTURE: The fibrous texture is due to long chains of silica. The chains are very loosely bound to each other and may be easily cleaved/shredded apart.

APPLICATIONS: Asbestos is used wherever a cheap, fire resistant material is desired. These places include building insulation and fire resistant clothing. Due to the health hazard, however, these traditional uses of asbestos are being limited or eliminated. Crocidolite has the highest mechanical strength and is used to reinforce cement in high pressure applications.

COMMENTS: Asbestos is the generic name given to fibrous filaments of silica based ceramic. The most commonly used of these are chrysotile which is fine, smooth and flexible; amosite which is straight and brittle; crocidolite which is also straight but not so brittle, and anthophyllite which is fragmented and brittle.

ASTATINE, At

MELTING TEMPERATURE: 302^0 C

BOILING TEMPERATURE: 337^0 C

THERMAL CONDUCTIVITY: 1.7 W/mK

CHEMICAL STABILITY: highly reactive

TOXICITY: Astatine is a highly radioactive element with a half life of about eight hours. It is similar to iodine in that it collects in the thyroid gland on contact.

APPLICATIONS: It has been used as a tracer especially in medical research.

BARIUM OXIDE, BaO

MELTING TEMPERATURE: 1920° C

BOILING TEMPERATURE: about 2000° C

MINIMUM MELTING POINT OF BINARY SYSTEM: 1425° C with alumina; 1141° C with beryllium oxide; 1370° C with silica
SPECIFIC GRAVITY: 5.72

SPECIFIC HEAT: 0.3 kJ/kgK at 20° C, 0.3 kJ/kgK at 1100° C

COEFFICIENT OF LINEAR EXPANSION: 7.5X10^{-6}/K
HARDNESS: 3.3 Mohs
MODULUS OF ELASTICITY: 58,000 MPa

ELECTRICAL RESISTIVITY: 10^6 Ohm.cm at 300° C

MAGNETIC SUSCEPTIBILITY: -3.7X10^{-7} m^3/kg
REFRACTIVE INDEX: 1.93
COLOR: clear to white
SOLUBILITY IN WATER: 3.48 g/100ml in cold water, 90.8 g/100ml in hot water
CHEMICAL STABILITY; RESISTANCE TO:
    SOLVENTS: Barium oxide reacts readily with water.
    GASES: It reacts on heating with air, carbon dioxide and sulfur dioxide.
TOXICITY: Contact exposure to skin and eyes causes irritation and ingestion can cause death.
STRUCTURE: cubic closed packed, similar to rock salt
APPLICATIONS: The most common use of barium oxide is as an additive to commercial glass to increase the index of refraction.

BERYLLIUM CARBIDE, Be$_2$C

DECOMPOSITION TEMPERATURE: about 2100° C
SPECIFIC GRAVITY: 1.9

SPECIFIC HEAT: 1.4 kJ/kgK at 20° C, 2.1 kJ/kgK at 1220° C

THERMAL CONDUCTIVITY: 26 W/mK at 20° C, 16 W/mK at 1220° C

COEFFICIENT OF LINEAR EXPANSION: 10^{-5}/K for 22° to 600° C HARDNESS: 3000 Knoop

MODULUS OF ELASTICITY: 386,000 MPa at 20° C, 262,000 MPa at 1220° C

FLEXURAL STRENGTH: 310 MPa at 20° C, 207 MPa at 1220° C
COMPRESSIVE STRENGTH: 760 MPa
POISSON'S RATIO: 0.1
COLOR: yellow
SOLUBILITY IN WATER: dilute
CHEMICAL STABILITY; RESISTANCE TO:
    GASES: It reacts with water vapor even at low temperature although slowly

below 200° C. Beryllium carbide also reacts with oxygen above 1250° C, and it reacts with nitrogen above 1000° C as well as ammonia above 750° C.

TOXICITY: This material is both a serious health hazard and a moderate fire hazard when in the powdered form. When inhaled it causes berylliosis which is a progressive lung ailment culminating in death.

STRUCTURE: hexagonal

APPLICATIONS: Beryllium carbide is used in exotic applications where a strong light material is desirable. It is expensive due to the health hazards processing entails.

BERYLLIUM OXIDE, BERYLLIA, BROMELLITE, BeO

MAXIMUM CONTINUOUS SERVICE TEMPERATURE: 2000° C phase change to tetragonal produces cracking; 1900° C in oxidizing atmosphere

MELTING TEMPERATURE: 2570° C

MINIMUM MELTING POINT OF BINARY SYSTEM: 1835° C with alumina; 1384° C with calcia; 1855° C with magnesium oxide; 1670° C with silica; 2150° C with thorium oxide; 1700° C with titanium oxide; 2145° C with zirconium oxide; 1141° C with barium oxide; 1302° C with strontium oxide

SPECIFIC GRAVITY: 3.0 for alpha, 2.69 for beta

SPECIFIC HEAT: 0.24 kJ/kgK at 20° C, 0.49 kJ/kgK at 1000° C and 0.52 kJ/kgK at 1500°-1800° C

THERMAL CONDUCTIVITY: 250 W/mK at 20° C, 28.7 W/mK at 1000° C and 18 W/mK at 1500° C, 1.7 W/mK at 1850° C

COEFFICIENT OF LINEAR EXPANSION: $9X10^{-6}$/K for 20° to 800° C

THERMAL SHOCK: poor resistance

HARDNESS: 1000 Knoop

MODULUS OF ELASTICITY: 386,000 MPa at 20° C, 345,000 MPa at 1220° C; 41,000 MPa for whisker

TENSILE STRENGTH: 97 MPa at 20° C, 70 MPa at 1650° C, 55 MPa at 2100° C; 14,000 MPa for whisker

RUPTURE MODULUS: 200-275 MPa at 20° C, 380 MPa at 800° C (about the maximum), 175 MPa at 1450° C

FLEXURAL STRENGTH: 200 MPa at 20° C, 275 MPa at 1220° C

SHEAR STRENGTH: 152,000 MPa at 20° C, 124,000 MPa at 1550° C

COMPRESSIVE STRENGTH: 1,380 MPa at 20° C, 690 MPa at 500° C

CREEP RATE: 0.0003/hr at 1300° C and a 12 MPa load; 0.005/hr at 2050° C and a 21 MPa load

POISSON'S RATIO: 0.34 for 20° to 1000° C

ELECTRICAL RESISTIVITY: 10^{17} Ohm.cm at 20° C, 10^{15} ohm.cm at 300° C,

5X10^{13} at 500° C, 1.5X10^{10} Ohm.cm at 700° C, 7X10^7 Ohm.cm at 1000° C
DIELECTRIC STRENGTH: 10.2 kv/mm
DIELECTRIC CONSTANT: 6 at 1 MHz to 1 GHz.
DISSIPATION FACTOR: 0.0005 at 1 MHz.; 0.001 at 1 GHz.
MAGNETIC SUSCEPTIBILITY: -1.5X10^{-7} m^3/kg
REFRACTIVE INDEX: 1.73
COLOR: white
SOLUBILITY IN WATER: 0.00002 g/100ml
DIFFUSION COEFFICIENT: 10^{-11} cm.cm./sec at 1600° C, 10^{-9} cm.cm/sec at 2000° C for oxygen; 10^{-11} cm.cm/sec at 1500° C,
8X10^{-10} cm.cm/sec at 2000° C for beryllium
CHEMICAL STABILITY; RESISTANCE TO:
ACIDS: Beryillia is not too resistant to acids, but generally better than Magnesia.
BASES: not resistant
GASES: It reacts with water vapor which becomes quite rapid above 1000° C. It also reacts with sulphur containing gases.
METALS: Beryillia has good resistance to even the most corrosive liquid metals. It also shows great stability in the presence of graphite.
TOXICITY: All beryllium compounds are toxic in the powdered form. If they are inhaled and they cause scar tissue growth around the powder in the lungs.
RESPONSE TO IRRADIATION: 0.51 neutron cross-section/cm. scattering, 0.715 effective scattering; macroscopic growth with accompaning grain boundary separation begins at 10^{20} nvt and is approximately linear with irradiation dose. Damage can be minimized by using fine ground ceramic and increasing the temperature. Often the annealing effect of the higher temperature can completely offset the irradiation effect.
STRUCTURE: hexagonal for alpha form, tetragonal for beta form. The alpha form transforms to the beta phase at about 2100° C.
APPLICATIONS: It is used to make crucibles and also as a additive to electrical porcelains. The special handling required to protect workers greatly adds to the cost of all products.

BORON, B

MELTING TEMPERATURE: 2300° C
BOILING TEMPERATURE: 3658° C
SPECIFIC GRAVITY: 2.34
SPECIFIC HEAT: 1.03 kJ/kgK
THERMAL CONDUCTIVITY: 27.0 W/mK
COEFFICIENT OF LINEAR THERMAL EXPANSION: 8.3X10^{-6}/K
HARDNESS: 49,000 Vickers
ELECTRICAL RESISTIVITY: 1.8X10^6 Ohm.cm

MAGNETIC SUSCEPTIBILITY: -9.8×10^{-11} m^3/kg
SOLUBILITY IN WATER: none
CHEMICAL RESISTIVITY:
    ACIDS: Boron is resistant to both hydrofluoric and hydrochloric acids as well as most other acids.
    BASES: It is resistant to most bases.
    GASES: Boron shows good resistance to oxygen and other gases.
    SOLVENTS: Water and other solvents are not effective on boron.
COLOR: yellow
TOXICITY: Boron compounds may be carcinogenic, especially with repeated exposure.
STRUCTURE: Boron can exist as monoclinic, tetragonal, rhombohedral, orthorhombic and hexagonal.
APPLICATIONS: Boron is used in the regulation of nuclear reactors. It is also used as a semi-conductor and for structural sections of aircraft. In fiber optics and heat resistant glass it is an extensively used additive.

BORON CARBIDE, B$_4$C

MAXIMUM CONTINUOUS SERVICE TEMPERATURE: 600° C for an oxidizing atmosphere, 2000° C for a reducing or inert atmosphere

MELTING TEMPERATURE: 2450° C

BOILING TEMPERATURE: 3770° C
SPECIFIC GRAVITY: 2.51

SPECIFIC HEAT: 0.85 kJ/kgK at 20° C., 2.1 kJ/kgK at 1220° C

THERMAL CONDUCTIVITY: 2.7 W/mK at 25° C, 8.4 W/mK at 425° C

COEFFICIENT OF LINEAR EXPANSION: 4.5×10^{-6}/K for 20° to 800° C
THERMAL SHOCK: poor resistance
HARDNESS: 2750-2880 Knoop, 9.5 Mohs

MODULUS OF ELASTICITY: 450,000 MPa at 20° C, 260,000 MPa at 1220° C

TENSILE STRENGTH: 303 MPa at 25° C, 150 MPa at 980° C

RUPTURE MODULUS: 350 MPa at 25° C, 160 MPa at 1425° C

FLEXURAL STRENGTH: 310 MPa at 20° C, 200 MPa at 1220° C
SHEAR STRENGTH: 2,800 MPa
SHEAR MODULUS: 200,000 MPa
COMPRESSIVE STRENGTH: 28 MPa

FRACTURE TOUGHNESS: 3 MPa.m$^{1/2}$
ABRASION RESISTANCE: very good to excellent
WEAR RATE: 0.08-0.14 mm.mm/kg.km on steel
POISSON'S RATIO: 0.19

ELECTRICAL RESISTIVITY: 0.3 Ohm.cm at 20° C, 0.02 Ohm.cm at 500° C
COLOR: black

ACOUSTIC LONGITUDINAL WAVE VELOCITY: 11 km/s

ACOUSTIC LONGITUDINAL WAVE IMPEDANCE: 26.4 kg/m^{2}s

SOLUBILITY IN WATER: none

INFRARED ABSORPTION BAND: 9.5×10^{-6} m to 12.5×10^{-6} m

CHEMICAL STABILITY; RESISTANCE TO:

ACIDS: very good resistance

BASES: very good resistance

GASES: Boron carbide reacts readily with oxygen above 1000° C. It also reacts slightly with sulphur containing gases and hydrogen above 1200° C.

TOXICITY: The dust is a nuisance but it is not known to be harmful.

THERMAL NEUTRON CAPTURE CROSS-SECTION: 600 barn

STRUCTURE: hexagonal

APPLICATIONS: Boron carbide is an abrasive second only to diamond. It also finds uses where wear resistance is important. There are nuclear applications due to its efficiency in absorbing thermal neutrons.

BORON NITRIDE, PYROLYTIC BORON NITRIDE(PBN), BN

MAXIMUM CONTINUOUS SERVICE TEMPERATURE: 1200° C for an oxidizing atmosphere, 2200° C for a reducing or inert atmosphere, 2000° C for a vacuum

MELTING TEMPERATURE: 2300° C for boron nitride

SUBLIMATION TEMPERATURE: about 3000° C for pyrolytic boron nitride

SPECIFIC GRAVITY: 2.0 for boron nitride, 2.3 for pyrolytic boron nitride

SPECIFIC HEAT: 0.6 kJ/kgK at 20° C, 1.7 kJ/kgK at 1060° C for boron nitride

THERMAL CONDUCTIVITY: 17-29 W/mK at 20° C, 27 W/mK at 839° C, 20 W/mK at 1424° C, 18.5 W/mK at 1841° C for boron nitride; 29 W/mK over 200° to 1400° C for pyrolytic boron nitride

COEFFICIENT OF LINEAR EXPANSION: 0.8×10^{-6}-7.5×10^{-6}/K over 0° to 1000° C for boron nitride; 2.4×10^{-5}/K over 0° to 1100° C for pyrolytic boron nitride

THERMAL SHOCK: exceptional resistance for pyrolytic boron nitride; good resistance for boron nitride

HARDNESS: 230 Knoop for boron nitride

MODULUS OF ELASTICITY: 34-85 MPa at 20° C, 3-10 MPa at 700° C for boron nitride

TENSILE STRENGTH: 25 MPa at 20° C; 2 MPa at 1000° C, 45 MPa at 2400° C for boron nitride; 41 MPa at 20° C, 55 MPa at 1600° C, 110 MPa at 2200° C for pyrolytic boron nitride

RUPTURE MODULUS: 50-110 MPa at 20° C, 13-26 MPa at 700° C, 7-15 MPa at 1000° C

FLEXURAL STRENGTH: 100 MPa at 20^0 C, 48 MPa at 1350^0 C for boron nitride

SHEAR STRENGTH: 52 MPa for boron nitride

COMPRESSIVE STRENGTH: 235-310 MPa

POISSON'S RATIO: 0.23

ELECTRICAL RESISTIVITY: 1.7×10^{13} Ohm.cm at 25^0 C, 2.3×10^{10} Ohm.cm at 500^0 C, 10^6 Ohm.cm at 850^0 C, 10^5 Ohm.cm at 1050^0 C, 0.001 Ohm.cm at 2000^0 C for boron nitride

DIELECTRIC STRENGTH: 45 kv/mm.

DIELECTRIC CONSTANT: 4.2 at 1 MHz

DISSIPATION FACTOR: 0.003 at 1 MHz

COLOR: The color is white for boron nitride, but it can be clear if pressed to form pyrolytic boron nitride.

ACOUSTIC LONGITUDINAL WAVE VELOCITY: 5.03 km/s

ACOUSTIC SHEAR WAVE VELOCITY: 3.86 km/s

ACOUSTIC LONGITUDINAL WAVE IMPEDANCE: 9.88 kg/m^{2}s

SOLUBILITY IN WATER: none

CHEMICAL STABILITY; RESISTANCE TO:

ACIDS: Pyrolytic boron nitride is very resistant.

BASES: Pyrolytic boron nitride can be decomposed by molten bases.

GASES: It is reactive in water vapor above 220^0 C. It is slightly reactive with hydrogen chloride above 800^0 C to form boron chloride; reacts with carbon bearing gases above 2200^0 C to form boron carbide. However, it will not react with carbon monoxide below 800^0 C.

METALS: It is not readily attacked by liquid metals so it is often used as a container when high purity is desired. Boron nitride is also more stable than diamond at elevated temperatures relative to reactions with ferrous metals.

GLASSES: Boron nitride will not react with glasses and thus can be used as a container for many slags, salts and glasses.

ORGANIC SOLVENTS: no effect

TOXICITY: Only the dust is a danger because it is a slight irritant.

WEAKNESSES: It is expensive to process and cannot be used in contact with graphite at high temperature.

STRUCTURE: The usual form is hexagonal with lamellar formate similar to graphite. The high temperature form is cubic with diamond structure. Because boron nitride can be processed by vapor decomposition it can also display the same structure as pyrolytic graphite.

APPLICATIONS: All forms are used as containment vessels for melting metals but the pyrolytic form is favored because of its purity. It finds use as a cutting tool and also as a very effective abrasive.

BROMINE, Br

MELTING TEMPERATURE: 266 K

BOILING TEMPERATURE: 58º C
SPECIFIC GRAVITY: 3.12
SPECIFIC HEAT: 0.29 KJ/KgK
THERMAL CONDUCTIVITY: 0.122 W/mK
MAGNETIC SUSCEPTIBILITY: $-4.44X10^{-9}$ m^3/kg
COLOR: dark red liquid
CHEMICAL STABILITY: highly reactive
TOXICITY: Bromine is highly irritating to the body, especially in the vapor form. It is also a moderate fire hazard due to its extreme reactivity.
STRUCTURE: orthorhombic
APPLICATIONS: Bromine makes an effective flame retardant. It is also used in water purification and dyes.

CADMIUM CHLORIDE, $CdCl_2$

MELTING TEMPERATURE: 568º C
BOILING TEMPERATURE: 960º C
SPECIFIC GRAVITY: 4.05
COLOR: clear
SOLUBILITY IN WATER: very slight
TOXICITY: Cadmium chloride is an irritant to eyes and mucus membranes. If ingested it could be fatal.
STRUCTURE: hexagonal
APPLICATIONS: It is photosensitive and finds many uses due to this attribute.

CADMIUM SULFIDE, GREENOCKITE, CdS

MELTING TEMPERATURE: 1750º C
SPECIFIC GRAVITY: 4.8
HARDNESS: 3.5 Mohs
MAGNETIC SUSCEPTIBILITY: $-6.28X10^{-7}$ m^3/kg
REFRACTIVE INDEX: 2.5
TRANSMISSION OF LIGHT: 10% or more through a 2mm thickness in the light range 0.55-16 microns
COLOR: yellow to orange
SOLUBILITY IN WATER: 0.0001 g/l00ml
TOXICITY: Cadmium sulfide is a carcinogen and otherwise nasty material which should be very carefully handled since any exposure to it can cause long term damage to the human body.
STRUCTURE: hexagonal
APPLICATIONS: It is used as a coloring agent in glazes. The usual color is yellow but in the presence of selenium it turns red.

CALCIUM CARBONATE, CALCITE, ARAGONITE, $CaCO_3$

MELTING TEMPERATURE: 1339^0 C
SPECIFIC GRAVITY: 2.93 for aragonite and 2.7 for calcite
SPECIFIC HEAT: 0.8 kJ/kgK
HARDNESS: 3 Mohs
MAGNETIC SUSCEPTIBILITY: -4.8×10^{-7} m^3/kg
REFRACTIVE INDEX: 1.5-1.7
BIREFRINGENCE: 0.17 for calcite
TRANSMISSION OF LIGHT: Calcite transmits at least 10% of incident light in the region 0.2-5.5 microns for a 2 mm thickness.
COLOR: clear
ACOUSTIC LONGITUDINAL WAVE VELOCITY: 3.8 km/s
ACOUSTIC LONGITUDINAL WAVE IMPEDANCE: 10.5 kg/m^2s
SOLUBILITY IN WATER: 0.0014 g/100ml for cold water and 0.002 g/100ml in hot water
TOXICITY: none
STRUCTURE: Aragonite has a rhombic crystalline structure. At 520^0 C it transforms into calcite which also has a rhombic structure.
APPLICATIONS: Calcium carbonate is a common food supplement and additive. It also finds use as a specialty optical window.

CALCIUM FLUORIDE, FLUORITE, CaF_2

MELTING TEMPERATURE: 1423^0 C
SPECIFIC GRAVITY: 3.2
SPECIFIC HEAT: 0.8 kJ/kgK
THERMAL CONDUCTIVITY: 0.8 W/mK
COEFFICIENT OF LINEAR EXPANSION: 10.1×10^{-6}/K
THERMAL SHOCK: poor resistance
HARDNESS: 4 Mohs; 200 Knoop
MODULUS OF ELASTICITY: 83,000-110,000 MPa
RUPTURE MODULUS: 37 MPa
FLEXURAL STRENGTH: 62-90 MPa
FRACTURE TOUGHNESS: 0.4-0.6 $MPa.m^{1/2}$
POISSON'S RATIO: 0.28
DIELECTRIC CONSTANT: 6.8
MAGNETIC SUSCEPTIBILITY: -2.52×10^{-7} m^3/kg
REFRACTIVE INDEX: 1.4
TRANSMISSION OF LIGHT: 10% for the range 1.13-11.8 microns and a 2 mm thickness
COLOR: clear
SOLUBILITY IN WATER: 0.0016 g/100ml at 18^0 C, 0.0017 g/100ml at 26^0 C
CHEMICAL STABILITY; RESISTANCE TO:

ACIDS: very resistant
BASES: very resistant
TOXICITY: It gives off harmful fluorine fumes which produce pinholes when it is used in a glaze.
STRUCTURE: cubic
APPLICATIONS: It is an excellent flux but tends to react with other components in a glaze. Many new applications in areas like lasers are being found.

CALCIUM OXIDE, CALCIA, LIME, CaO

MELTING/DECOMPOSITION TEMPERATURE: 2610° C

BOILING TEMPERATURE: 2850° C

MINIMUM MELTING POINT OF BINARY SYSTEM: 1360° C with alumina; 1384° C with beryllium oxide; 2300° C with magnesium oxide; 1436° C with silica; 2300° C with thoria; 1420° C with titania; 2140° C with zirconia;
SPECIFIC GRAVITY: 3.25-3.38

SPECIFIC HEAT: 0.76 kJ/kgK at 20° C, 0.95 kJ/kgK at 1100° C

THERMAL CONDUCTIVITY: 29 W/mK at 20° C, 17 W/mK at 1100° C

COEFFICIENT OF LINEAR EXPANSION: 5×10^{-6}/K
THERMAL SHOCK: moderate resistance
HARDNESS: 550 Knoop; 4.5 Mohs
MODULUS OF ELASTICITY: 114,000 MPa
TENSILE STRENGTH: 78 MPa

FLEXURAL STRENGTH: 103 MPa at 20° C, 69 MPa at 1200° C

MAGNETIC SUSCEPTIBILITY: -1.88×10^{-7} m^3/kg
REFRACTIVE INDEX: 1.84
COLOR: clear

SOLUBILITY IN WATER: 0.13 g/100ml at 10° C, 0.07 g/100ml at 80° C

DIFFUSION COEFFICIENT: 10^{-14} cm.cm/sec at 1000° C, 10^{-10} cm.cm/sec at 1600° C for calcium
CHEMICAL STABILITY; RESISTANCE TO:
  GASES: Calcia reacts on heating with carbon dioxide and sulfur dioxide.
  METALS: It has good resistance to lithium vapor up to 2000° C.
    SOLVENTS: Calcia reacts with water.
TOXICITY: Calcia is used as a food additive and is nontoxic but should not be inhaled due to its nuisance factor.
STRUCTURE: cubic close packed similar to rock salt
APPLICATIONS: This is a primary raw ingredient in concrete and plaster.

CARBON, GRAPHITE, PYROLYTIC GRAPHITE, DIAMOND, C

MAXIMUM CONTINUOUS SERVICE TEMPERATURE: at 2200° C pyrolytic graphite begins to convert to natural graphite structure; at 1000° C diamond begins to transform into graphite, this transformation becomes rapid at about 1750° C; 450° C for graphite in oxidizing atmosphere; 550° C for vitreous carbon in an oxidizing atmosphere; 2500° C for both graphite and vitreous carbon in a reducing or inert atmosphere

MELTING TEMPERATURE: 3652° C for graphite and vitreous carbon; about 3550° C for diamond

BOILING TEMPERATURE: 4827° C for graphite and vitreous carbon

ANNEALING TEMPERATURE: 1500° C for all forms

SPECIFIC GRAVITY: 1.44-2.265 for graphite; 3.52 for diamond; 1.8-2.1 for vitreous carbon

SPECIFIC HEAT: 1.0 kJ/kgK at 20° C, 1.4 kJ/kgK at 225° C, 1.3 kJ/kgK at 725° C, 2.1 kJ/kgK at 1725° C, 2.2 kJ/kgK at 2725° C, 2.8 kJ/kgK at 3225° C for both graphite and charcoal; 0.5 kJ/kgK for diamond

THERMAL CONDUCTIVITY: 6.3 W/mK at 8 K, 58.7 W/mK at 83 K, 293 W/mK at 98 K, 96.4 W/mK at 20° C, 18 W/mK at 100° C and 6.3 W/mK at 1000° C for graphite; 11,000 W/mK at 100 K, 60 W/mK at 200 K, 26 W/mK for diamond at 20° C; 139 W/mK at 96° C, 87 W/mK at 750° C for pyrolytic graphite

COEFFICIENT OF LINEAR EXPANSION: $1-1.9X10^{-6}$/K for graphite; $4.6X10^{-6}$/K at 20° to 100° C, $6.7X10^{-6}$/K at 100° to 200° C, $8.6X10^{-6}$/K at 200° to 300° C, $9.8X10^{-6}$/K at 300° to 400° C, $10.7X10^{-6}$/K at 400° to 500° C, $12X10^{-6}$/K at 500° to 700° C for diamond

THERMAL SHOCK: excellent resistance

HARDNESS: 7000 Knoop and 10 Mohs for diamond; 1-2 Mohs for graphite

MODULUS OF ELASTICITY: 4,400-8,000 MPa for graphite; 900,000 MPa for diamond; 1,000,000 MPa for graphite whisker; 20,700 MPa at 20° to 360° C, 13,800 MPa at 800° to 1200° C for pyrolytic graphite

TENSILE STRENGTH: 3-9 MPa at 20° C, 4 MPa at 1650° C, 14 MPa at 2420° C, 8 MPa at 2750° C for graphite; 19 MPa at 20° C, 25 MPa at 1650° C, 35 MPa at 2400° C, 21 MPa at 2750° C for pyrolytic graphite; 29 MPa. for diamond

RUPTURE STRENGTH: 200 MPa for graphite

FLEXURAL STRENGTH: 6-18 MPa for graphite; 200-500 MPa for diamond; 172 MPa at 0° C, 220 MPa at 2000° C for pyrolytic graphite

COMPRESSIVE STRENGTH: 41-44 MPa for graphite; 2,000 MPa for diamond; 345 MPa at 0° C, 366 MPa at 700° C for pyrolytic graphite

FRACTURE TOUGHNESS: 1 MPa.m$^{1/2}$ for graphite

POISSON'S RATIO: 0.07-0.22 for graphite

COEFFICIENT OF FRICTION: 0.1 for diamond on diamond, 0.1-0.15 for diamond on metal; 0.1 for graphite on graphite or steel

ABRASION RESISTANCE: Diamond is the hardest naturally occurring substance and is only effectively abraded by itself. When abrasion occurs between diamonds the residue is graphite which seems to form due to the heat generated. Graphite is soft and easily abraded. Pyrolytic graphite is slightly less abradable than normal graphite.

WEAR RATE: 0.001-0.002 mm.mm/kg.km on steel, 0.00013 mm.mm/kg.km on gray cast iron for diamond

ELECTRICAL RESISTIVITY: 0.00085-0.00116 Ohm.cm at 20° C, 0.0005 Ohm.cm at 500° C for graphite; 10^{16} Ohm.cm for diamond but it drops off rapidly with defects or impurities

DIELECTRIC CONSTANT: 5.7 at 1 MHz for diamond

DISSIPATION FACTOR: 0.0002 at 1 MHz for diamond

BAND GAP: 5.45 for diamond

MAGNETIC SUSCEPTIBILITY: -7.6×10^{-8} m^3/kg for graphite; -7.4×10^{-8} m^3/kg for diamond

SATURATED ELECTRON VELOCITY: 2.7×10^7 cm/s for diamond

BREAKDOWN: 10^7 V/cm for diamond

CARRIER MOBILITY:

ELECTRON: 2,200 cm^2/Vs for diamond

HOLE: 1,600 cm^2/Vs for diamond

ABSORPTION EDGE: 0.22 microns for diamond

REFRACTIVE INDEX: 2.42 for diamond

COLOR: Graphite is gray to black, carbon is black and diamond is clear to cloudy.

ACOUSTIC LONGITUDINAL WAVE VELOCITY: 4.26 km/s for vitreous graphite, 3.31 km/s for pyrolytic graphite

ACOUSTIC SHEAR WAVE VELOCITY: 2.68 km/s for vitreous graphite

ACOUSTIC LONGITUDINAL WAVE IMPEDANCE: 6.26 kg/m^2s for vitreous graphite, 7.31 kg/m^2s for pyrolytic graphite

SOLUBILITY IN WATER: none

CHEMICAL STABILITY; RESISTANCE TO:

ACIDS: There is no effect below 450° C except on graphite by nitric acid.

BASES: Carbon is not effected below 450° C.

GASES: Graphite reacts readily with air and oxygen above 500° C, water vapor above 700° C and carbon dioxide above 900° C. It also reacts slightly with hydrogen above 800° C. Vitreous carbon reacts slightly with oxygen and air above 500° C as well as water vapor and carbon dioxide above 1100° C. Diamond reacts with oxygen above 600° C and sodium nitrate above 450° C.

METALS: It is resistant to all metals in the solid or liquid state except those that form carbides.

SOLVENTS: Carbon resistant to all organic solvents.

GLASSES: Carbon is resistant to glasses and slags.

TOXICITY: In the powdered form carbon is carcinogenic both by contact to skin and inhaling into the lungs. The skin reaction seems to be fairly complex and involves the presence of other substances.

RESPONSE TO IRRADIATION: graphite expands in the long axis of the hexagonal unit cell. It also decreases in both number and size of pores. For all forms, except diamond, restructuring can take place resulting in a material that is so cross-linked it forms one large molecule. The neutron cross section/cm scattering is 0.409 and the effective scattering is 0.404 for graphite.

RADIOACTIVE EMITTANCE: 0.8

STRUCTURE: The structure is hexagonal for graphite, aligned hexagonal for pyrolytic graphite and face centered cubic for diamond. Most forms of carbon are amorphous.

ADDITIVES: The addition of about 0.5% boron to graphite increases the electrical conductivity such that it behaves as a metal. The same amount of boron added to pyrolytic graphite increases its strength. Small impurities in diamond result in a beautiful array of colors.

APPLICATIONS: Pyrolytic graphite retains its strength to very high temperatures and thus is used when a high strength, high purity but expensive material is needed such as in rocket motor nozzles. Vitreous carbon is a fairly dense amorphous structure with microporosity. This closed form makes it inert and thus applicable to human contact. It is currently being used as a dental implant material. Graphite is used as a refractory and as a lubricant. Diamonds of all grades are used as cutting tools.

CERIUM OXIDE, CERIA, CeO_2

MELTING TEMPERATURE: approximately 2600^0 C
SPECIFIC GRAVITY: 7.13
SPECIFIC HEAT: 0.36 kJ/kgK
THERMAL CONDUCTIVITY: 12.1 W/mK
COEFFICIENT OF LINEAR EXPANSION: 10^{-5}/K
THERMAL SHOCK RESISTANCE: poor
HARDNESS: 6 Mohs
TENSILE STRENGTH: 88 MPa
SHEAR STRENGTH: 110 MPa
COMPRESSIVE STRENGTH: 1386 MPa
ELECTRICAL RESISTIVITY: 10^9 Ohm.cm at 20^0 C, 10^5 Ohm.cm at 800^0 C, 10^3 Ohm.cm at 1200^0 C
DIELECTRIC CONSTANT: 15 at 1 MHz
DISSIPATION FACTOR: 0.01 at 1 MHz
MAGNETIC SUSCEPTIBILITY: $3.3X10^{-7}$ m^3/kg
COLOR: brown
SOLUBILITY IN WATER: none
TOXICITY: The ingestion of cerium oxide results in psychological disorientation and possibly death. It also reacts readily with water and can result in a fire hazard.
STRUCTURE: cubic

APPLICATIONS: Its uses have been primarily magnetic.

CESIUM IODIDE, CsI

MELTING TEMPERATURE: 626° C
BOILING TEMPERATURE: 1280° C
SPECIFIC GRAVITY: 4.5
SPECIFIC HEAT: 0.2 kJ/kgK
HARDNESS: below testing range for most procedures
MAGNETIC SUSCEPTIBILITY: -1.04×10^{-6} m^3/kg
REFRACTIVE INDEX: 1.8
TRANSMISSION OF LIGHT: At least 10% of incident light is transmitted through a 2 mm thickness in the region of 0.25-70 microns.
COLOR: white
SOLUBILITY IN WATER: 44 g/100ml in cold water and 160 g/100ml in hot water
TOXICITY: It is non-toxic while in the solid form but very irritating if dissociated. On heating the material emits highly toxic fumes of iodine and iodine compounds.
STRUCTURE: rhombic
APPLICATIONS: Its very wide light transmission window makes it useful for various types of sighting devices.

CHLORINE, Cl

MELTING TEMPERATURE: 172 K
BOILING TEMPERATURE: 238 K
SPECIFIC GRAVITY: 0.0032
SPECIFIC HEAT: 0.49 KJ/KgK
THERMAL CONDUCTIVITY: 0.00641 W/mK at 233 K, 0.00727 W/mK at 255 K, 0.00970 W/mK at 49° C
DIELECTRIC CONSTANT: 2.1 at 233 K, 1.91 at 14 C, 1.7 at 77 C, 1.5 at 142 C
MAGNETIC SUSCEPTIBILITY: -7.2×10^{-9} m^3/kg
COLOR: yellow gas
CHEMICAL STABILITY: highly reactive
TOXICITY: Like bromine, it is highly irritating and a moderate fire hazard.
STRUCTURE: Chlorine is tetragonal until 100 K when it transforms into orthorhombic.
APPLICATIONS: Water purification is a world wide use for chlorine.

CHROMIUM CARBIDE, Cr_3C_2

MAXIMUM CONTINUOUS SERVICE TEMPERATURE: 1220° C in air, 2200°

C in vacuum

MELTING TEMPERATURE: 1890° C

BOILING TEMPERATURE: 3800° C

SPECIFIC GRAVITY: 6.68

SPECIFIC HEAT: 0.52 kJ/kgK at 20° C, 0.88 kJ/kgK at 1200° C

THERMAL CONDUCTIVITY: 19 W/mK

COEFFICIENT OF LINEAR EXPANSION: 9.9X10^{-6}/K for 22° to 1000° C

HARDNESS: 1800 Knoop

MODULUS OF ELASTICITY: 400,000 MPa

RUPTURE STRENGTH: 49 MPa

FLEXURAL STRENGTH: 275 MPa at 20° C, 550 MPa at 1200° C

COMPRESSIVE STRENGTH: 1100 MPa at 20° C, 410 MPa at 1400° C

ELECTRICAL RESISTIVITY: 1,270 Ohm.cm at 350° C, 75 Ohm.cm at 750° C

COLOR: gray

SOLUBILITY IN WATER: none

CHEMICAL STABILITY; RESISTANCE TO:

ACIDS: excellent

BASES: resistant up to 700° C

TOXICITY: It is carcinogenic on contact and causes ulcerations which heal slowly.

STRUCTURE: orthorhombic

APPLICATIONS: It is used as filters and pump bearings in the chemical industry.

CHROMIUM OXIDE, CHROMIC OXIDE, CHROMIA, CHROMITE, Cr_2O_3

MAXIMUM CONTINUOUS SERVICE TEMPERATURE: 880° C in a reducing atmosphere

MELTING TEMPERATURE: 2,435° C

BOILING TEMPERATURE: 4000° C

MINIMUM MELTING POINT FOR BINARY SYSTEM: 2030° C for alumina; 2340° C for magnesium oxide; 1710° C for silica

SPECIFIC GRAVITY: 5.21

SPECIFIC HEAT: 0.72 kJ/kgK at 20° C, 0.80 kJ/kgK at 1100° C

THERMAL CONDUCTIVITY: 10-33 W/mK

COEFFICIENT OF LINEAR EXPANSION: 2.3X10^{-6}/K for 350° to 1200° C

HARDNESS: 9 Mohs; 3000 Knoop

MODULUS OF ELASTICITY: 103,000 MPa

RUPTURE STRENGTH: 262 MPa

FLEXURAL STRENGTH: 70 MPa

FRACTURE TOUGHNESS: 3.9 MPa.m$^{1/2}$

CREEP RATE: 1.4X10^{-5}/hr for 0.7 MPa load, 7.2X10^{-4}/hr for a 70 MPa load, both are at 1200° C

ELECTRICAL RESISTIVITY: 1,000 Ohm.cm at 350° C, 0.2 Ohm.cm at 1200° C

MAGNETIC SUSCEPTIBILITY: 2.51×10^{-6} m^3/kg

NEEL TEMPERATURE: 307 K

CURIE-WEISS TEMPERATURE: 212° C

REFRACTIVE INDEX: 2.55

COLOR: green

SOLUBILITY IN WATER: none

DIFFUSION COEFFICIENT: 10^{-14} cm.cm/sec at 1200° C, 5×10^{-12} cm.cm/sec at 1500° C for oxygen through chromium oxide; 10^{-11} cm.cm/sec at 1000° C, 7×10^{-9} cm.cm/sec at 1500° C for chromium through chromium oxide

CHEMICAL STABILITY; RESISTANCE TO:

ACIDS: reacts readily

BASES: reacts readily

GASES: reacts readily with hydrogen above 860° C and oxygen above 1000° C

METALS: poor resistance to liquid form

GLASSES: very resistant to glasses and slags

TOXICITY: It is carcinogenic on contact and causes ulcerations which heal slowly.

IRRADIATION EMITTANCE: 0.69

EFFECTIVE MOVEMENT PER ATOM: 3.73 erg/gauss

MAGNETIC MOVEMENT PER ATOM: 3.0 erg/gauss

STRUCTURE: hexagonal closed packed, similar to corundum

APPLICATIONS: It is the most stable green pigment known so it finds extensive decorative uses. In conjunction with tin, calcium or silica it forms pink, with zinc it produces brown and with high amounts of lead it produces yellow in glazes. Chromium oxide is also used as a containment vessel for slags.

CONCRETE

MAXIMUM CONTINUOUS SERVICE TEMPERATURE: 600° C for Portland cement based concrete, up to 1750° C for refractory concrete

MELTING TEMPERATURE: Since this is a composite some localized melting is possible while the material is still functional, thus there is no identifiable melting temperature.

SPECIFIC GRAVITY: 7.86 for steel reinforced concrete; 2.7 for glass reinforced concrete; 0.91 for polypropylene reinforced concrete; 1.7-2.5 for Portland cement; generally as specific gravity is increased so is durability and strength

THERMAL CONDUCTIVITY: 1.6 W/mK for Portland concrete; 0.4-2.0 W/mK for sulphur concrete

COEFFICIENT OF LINEAR EXPANSION: 11×10^{-6}/K for Portland concrete; 18×10^{-6}/K for Portland cement; 8×10^{-6}-35×10^{-6}/K for sulphur concrete

THERMAL SHOCK RESISTANCE: excellent for refractory grades, good other-wise

ELONGATION: 3.5 % for steel or glass reinforced concrete; 3% for polypropy-lene reinforced concrete; 0.005-0.015% for Portland cement

MODULUS OF ELASTICITY: 200,000 MPa for steel reinforced concrete; 80,000 MPa for glass reinforced concrete; 3,000-15,000 MPa for polypropylene reinforced concrete; 30,000-40,000 MPa for Portland cement; 28,000 MPa for Portland con-crete; 20,000-45,000 MPa for sulphur concrete

TENSILE STRENGTH: 34.5 MPa for ferrocement; 1,000-3,000 MPa for steel reinforced concrete; 2500 MPa for glass reinforced concrete; 400 MPa for polypropylene reinforced concrete; 1-4 MPa for Portland cement; 3.5 MPa for Portland concrete; 2.8-8.3 MPa for sulphur concrete. In all cases the strength is dependent on the tricalcium silicate content. As this increases so does the strength.

RUPTURE MODULUS: 55.1 MPa for ferrocement; 3.7 MPa for Portland con-crete; 3.4-10.4 MPa for sulphur concrete

FLEXURAL STRENGTH: 7 MPa for Portland cement

COMPRESSIVE STRENGTH: 27.6-68.9 MPa for ferrocement; 35 MPa for Portland concrete; 28-70 MPa for sulphur concrete; 40 MPa for Portland cement

ABRASION RESISTANCE: poor

FRACTURE ENERGY: 15 J/m.m for Portland cement

ACOUSTIC LONGITUDINAL WAVE VELOCITY: 3.1 km/s

ACOUSTIC LONGITUDINAL WAVE IMPEDANCE: 8.0 kg/m^{2}s

CHEMICAL STABILITY; RESISTANCE TO:

ACIDS: Ammonium sulfate will attack all forms of cement including the sul-fate resistant forms. The calcium hydroxide component of cured cement is suscep-tible to leaching by acidic water. Concrete is particularly susceptible to organic acids.

BASES: relatively resistant

SOLVENTS: It is non-reactive to sea water if the material is totally sub-merged.

OILS AND FATS: Vegetable oils are more active in attacking concrete than animal fats but both cause deterioration.

ADDITIVES: An aggregate of stone and sand is used as a strengthening agent. Water is used to make the slurry workable but excess water adds to porosity. Calcium sulphate is used to retard the instantaneous setting of the tricalcium alu-minate in cement. Sugar in concentrations of less than 1% is a good retarder to concrete setting, however, it can produce organic acids which attack the concrete. Calcium chloride is used to reduce setting time in concentrations up to 2%. It pro-duces a slight decrease in tensile strength, flexural strength and corrosion resist-ance. Naphthalene sulphonate is among a group of superplasticizers that can reduce the amount of water needed to mix into concrete by as much as 30%. The result is a more workable slurry with a greater final strength.

IMPURITIES: Magnesium oxide is usually present in cement. It does not react but goes into solution on mixing. Later it recrystallizes as periclase before trans-forming to magnesium hydroxide. Magnesium hydroxide usually forms after the concrete has set. Since its formation causes an increase in volume there is the pos-sibility of it cracking the concrete.

CHEMICAL COMPOSITION: The precise chemistry of cement is still in doubt but most of it is tricalcium silicate, dicalcium silicate, tricalcium aluminate and tetracalcium alumino-ferrite.

STRUCTURE: a composite with both free and chemically bond components.

APPLICATIONS: Ferrocement is used to make boat hulls, silos, tanks and roofs. Natural fiber reinforced concrete is used for low cost houses where durability is less of a factor than weight and cost. Where would the building industry be without it?

COPPER BROMIDE, Cu_2Br_2

MELTING TEMPERATURE: 504° C

BOILING TEMPERATURE: 1345° C

SPECIFIC GRAVITY: 4.72

REFRACTIVE INDEX: 2.12

COLOR: white

SOLUBILITY IN WATER: slight

TOXICITY: With prolonged exposure copper bromide produces a rash, depression and mental deterioration.

STRUCTURE: cubic

APPLICATIONS: It is photosensitive and finds any uses due to this attribute.

COPPER CHLORIDE, NANTOKITE, Cu_2Cl_2

MELTING TEMPERATURE: 422° C

BOILING TEMPERATURE: 1366° C

SPECIFIC GRAVITY: 3.53

REFRACTIVE INDEX: 1.97

COLOR: white

SOLUBILITY IN WATER: 0.006 g/100ml

TOXICITY: Copper chloride is an irritant to eyes and mucous membranes.

STRUCTURE: cubic

APPLICATIONS: It is photosensitive and finds any uses due to this attribute.

COPPER IODIDE, MARSHITE, Cu_2I_2

MELTING TEMPERATURE: 605° C

BOILING TEMPERATURE: 1290° C

SPECIFIC GRAVITY: 5.63

REFRACTIVE INDEX: 2.34

COLOR: white

SOLUBILITY IN WATER: 0.0008 g/100ml

TOXICITY: Copper iodide is an irritant to eyes and mucous membranes.
STRUCTURE: cubic
APPLICATIONS: It is photosensitive and finds many uses due to this attribute.

EARTHENWARE, STONEWARE

MAXIMUM CONTINUOUS SERVICE TEMPERATURE: 1700^0 C
MELTING TEMPERATURE: due to the complex nature of earthenware localized
melting of some components may occur without rendering the material useless.
SPECIFIC GRAVITY: 2.1-2.6
SPECIFIC HEAT: 0.2 kJ/kgK

THERMAL CONDUCTIVITY: 0.4 W/mK from 20^0 to 1200^0 C

COEFFICIENT OF LINEAR EXPANSION: $4X10^{-6}$/K
MODULUS OF ELASTICITY: 8,280 MPa
TENSILE STRENGTH: 14 MPa
RUPTURE STRENGTH: 48-62 MPa
SHEAR STRENGTH: 70 MPa
COMPRESSIVE STRENGTH: 520 MPa
ABRASION RESISTANCE: This is greatly enhanced when a glaze is used. The
abrasion resistance becomes that of the glaze which is essentially that of silica.
Otherwise the abrasion resistance is poor.
ACOUSTIC LONGITUDINAL WAVE VELOCITY: 4.3 km/s

ACOUSTIC LONGITUDINAL WAVE IMPEDANCE: 7.4 kg/m^2s
WATER ABSORPTION: 0-2% of volume
CHEMICAL STABILITY; RESISTANCE TO:
    ACIDS: resistant to all except hydrofluoric acid
    BASES: resistant to high temperature
    ORGANIC SOLVENTS: very good
    GASES: very good
TOXICITY: In general it is non-toxic unless ingested. The dust is an irritant and
trace impurities can be of concern.
STRUCTURE: a complex crystalline and amorphous material made from natural
clay with a high silica content.
APPLICATIONS: Earthenware has been used throughout history for brick, tile,
clay pipe and pottery. New applications are being found in the chemical industry.

FERRIC ALUMINOSILICATE, GARNET, YAG, $Al_2O_3.3FeO\cdot3SiO_2$

SPECIFIC GRAVITY: 3.5-4.2
HARDNESS: 1360 Knoop; 6.5-7.5 Mohs
WEAR RATE: 0.054-0.066 mm.mm/kg.km on steel
REFRACTIVE INDEX: 1.8
COLOR: dark red
CHEMICAL STABILITY; RESISTANCE TO:

GASES: hydrogen reacts with garnet above 1100° C to form volatile products
STRUCTURE: Garnet is basically a solid solution of alumina, silica and iron oxide which produces a complex cubic crystal. Its composition can vary widely. For the most part the properties are like those of silica.
APPLICATIONS: Garnents have long been used for their decorative properties but commercial interests center on their abrasiveness. It makes an excellent sandpaper. Currently is is under consideration for bubble memory technology.

FERRIC OXIDE, HEMATITE, MAGNETITE, Fe_2O_3

MELTING TEMPERATURE: 1594° C for magnetite, 1562° C for hematite
SPECIFIC GRAVITY: 5.2
SPECIFIC HEAT: 0.6 kJ/kgK for hematite, 0.9 kJ/kgK for magnetite
HARDNESS: 6 Mohs
MAGNETIC SUSCEPTIBILITY: 3.76×10^{-5} m^3/kg at 760° C
CURIE TEMPERATURE: 620° C
REFRACTIVE INDEX: 2.4 for magnetite, 3.0 for hematite
COLOR: red to brown
SOLUBILITY IN WATER: none
TOXICITY: It is suspected to be carcinogenic especially in the powdered or vapor form. Caution should be used if the material is inhaled or ingested.
STRUCTURE: hematite is trigonal and magnetite is cubic
APPLICATIONS: A common component in clay, ferric oxide produces reds, browns or yellows in brick.

FERRITES

MELTING TEMPERATURE: variable depending on composition
SPECIFIC GRAVITY: 4.8
MODULUS OF ELASTICITY: 170,000 MPa
TENSILE STRENGTH: 215 MPa
POISSON'S RATIO: 0.3
FRACTURE TOUGHNESS: 2 MPa.m$^{1/2}$
FERROMAGNETIC: at low temperature
NEEL TEMPERATURE: $300\text{-}600^o$ C
TOXICITY: similar to the iron oxides
SATURATION MAGNETIZATION: 150-5000 Gauss
CURIE TEMPERATURE: $100\text{-}640^o$ C
RESONANCE LINE WIDTH: 10-1000 Oe
SPINWAVE LINE WIDTH: 1-20 Oe
DIELECTRIC CONSTANT: 12-18
DIELECTRIC LOSS TANGENT: 0.0001-0.002

SCALAR PERMEABILITY: approaches unity
MAGNETIC LOSS TANGENT: 0.001-1
REMNANT FLUX DENSITY: 50-4000 Gauss
COERCIVE FORCE: 0.2-3 Oe
GYROMAGNETIC RATIO: 2.8 MHz/Oe
STRUCTURE: The most common composition is $MO \cdot XFe_2O_3$ where M is usually Ba or Sr and X is a small number. $BaO \cdot 6Fe_2O_3$ forms hexagonal platelets.
APPLICATIONS: This is a general class of permanent magnet.

FLUORINE, F

MELTING TEMPERATURE: 53 K
BOILING TEMPERATURE: 85 K
SPECIFIC GRAVITY: 0.0017
SPECIFIC HEAT: 0.75 KJ/KgK
THERMAL CONDUCTIVITY: 0.00762 W/mK at 87 K, 0.0182 W/mK at 200 K, 0.0213 W/mK at 233 K, 0.0232 W/mK at 255 K, 0.0277 W/mK at 38° C, 0.0319 W/mK at 93° C
DIELECTRIC CONSTANT: 1.54 at 71 K
COLOR: pale yellow gas
CHEMICAL STABILITY: Fluorine is the most reactive of all elements.
TOXICITY: Because it is strongly reactive it is irritating to all living things.
HAZARD: Fluorine is a fire hazard and becomes dangerous if heated above ambient.
APPLICATIONS: As AlF_3 it is used in aluminum production.

GADOLINIUM OXIDE, GADOLINIA, Gd_2O_3

MELTING TEMPERATURE: 2330° C
SPECIFIC GRAVITY: 7.41
SPECIFIC HEAT: 0.26 kJ/kgK at 20° C, 0.34 kJ/kgK at 1200° C, 0.36 kJ/kgK at 1750° C
COEFFICIENT OF LINEAR EXPANSION: $2.5X10^{-6}$/K for 22° to 650° C, $2.9X10^{-6}$/K for 650° to 1500° C
HARDNESS: 380-550 Knoop
MODULUS OF ELASTICITY: 138,000 MPa at 20° C, 124,000 MPa at 1500° C
FLEXURAL STRENGTH: 117-138 MPa for 20° to 1400° C
SHEAR MODULUS: 55,000 MPa at 20° C, 48,000 MPa at 1500° C
POISSON'S RATIO: 0.276 at 20° C, 0.267 at 1000° C
COMPRESSIVE STRENGTH: 166-214 MPa
MAGNETIC SUSCEPTIBILITY: $6.28X10^{-4}$ m^3/kg

COLOR: white
SOLUBILITY IN WATER: very slight
CHEMICAL STABILITY; RESISTANCE TO:
GASES: reacts slowly with water vapor
TOXICITY: It is an irritant in the powdered form which should not be inhaled or ingested.
STRUCTURE: cubic
APPLICATIONS: Most of the application of this rare and expensive material are found in the magnetic area.

GALLIUM ARSENIDE, GaAs

MELTING TEMPERATURE: 1238° C
SPECIFIC GRAVITY: 5.32
SPECIFIC HEAT: 0.320 kJ/kgK
THERMAL CONDUCTIVITY: 46 W/mK
COEFFICIENT OF LINEAR THERMAL EXPANSION: $5.9X10^{-6}$/K
HARDNESS: 600 kg/mm^2
ELECTRICAL RESISTIVITY: 10^8 Ohm-cm
DIELECTRIC CONSTANT: 12.5
BAND GAP: 1.43 eV
SATURATED ELECTRON VELOCITY: 10^7 cm/s
CARRIER MOBILITY:

ELECTRON: 8,500 cm^2/Vs

HOLE: 400 cm^2/Vs

BREAKDOWN: $6X10^5$ V/cm
REFRACTIVE INDEX: 3.4
COLOR: gray
ABSORPTION EDGE: 0.85 microns
STRUCTURE: cubic
TOXICITY: Gallium arsenide is highly toxic and a known carcinogen.
APPLICATIONS: It is used as a semi-conductor and as a laser window.

GALLIUM NITRIDE, GaN

MELTING TEMPERATURE: 2830° C
SPECIFIC GRAVITY: 3.2
THERMAL CONDUCTIVITY: 490 W/mK
COEFFICIENT OF LINEAR THERMAL EXPANSION: $4.2X10^{-6}$/K
ELECTRICAL RESISTIVITY: Gallium nitride is superconductive below 5.85 K
DIELECTRIC CONSTANT: 9.8
ABSORPTION EDGE: 0.4 microns
BAND GAP: 3.02 eV

SATURATED ELECTRON VELOCITY: $2X10^7$ cm/s
CARRIER MOBILITY:

ELECTRONS: 370 cm^2/Vs

HOLE: 90 cm^2/Vs

BREAKDOWN: $3.2X10^6$ V/cm
REFRACTIVE INDEX: 2.7
COLOR: gray
SOLUBILITY IN WATER: none
STRUCTURE: compacted powder
TOXICITY: Gallium nitride is hazardous to people on contact.
APPLICATIONS: It is used as a semi-conductor.

GERMANIUM, Ge

MELTING TEMPERATURE: 937° C

BOILING TEMPERATURE: 2830° C
SPECIFIC GRAVITY: 5.35

SPECIFIC HEAT: 0.3 kJ/kgK at 20° C, 0.4 kJ/kgK at melting point

THERMAL CONDUCTIVITY: 50 W/mK at 20° C, 11 W/mK at 900° C

COEFFICIENT OF LINEAR THERMAL EXPANSION: $5X10^{-6}$/K

HARDNESS: 850 kg/mm^2
TENSILE STRENGTH: 70 MPa
MODULUS OF ELASTICITY: 130,000 MPa

SHEAR MODULUS: 67,000 MPa at 20° C, 61,000 MPa at 800° C
POISSON'S RATIO: 0.27
ELECTRICAL RESISTIVITY: 46 Ohm.cm
ENERGY GAP: 0.67 electron-volts

DIELECTRIC CONSTANT: 16.4 at 20° C, 17.6 at 200° C

MAGNETIC SUSCEPTIBILITY: $-1.328X10^{-9}$ m^3/kg
SOLUBILITY IN WATER: none
REFRACTIVE INDEX: 4
COLOR: gray
TOXICITY: none
HAZARD: It is a moderate fire hazard especially in the form of dust. If germanium is exposed to heat it can explode.
STRUCTURE: face centered cubic
APPLICATIONS: It is used as a semi-conductor as well as wide-angle lenses.

HAFNIUM CARBIDE, HfC

MELTING TEMPERATURE: 3890°

SPECIFIC GRAVITY: 12.2

SPECIFIC HEAT: 0.18 kJ/kgK at 20° C, 0.27 kJ/kgK at 1200° C

THERMAL CONDUCTIVITY: 11 W/mK at 20° C, 25 W/mK at 1200° C

COEFFICIENT OF LINEAR EXPANSION: 1.9×10^{-6}/K for 20° to 800° C, 2.8×10^{-6}/K for 800° to 2200° C

HARDNESS: 2500 Knoop at 20° C, 1000 Knoop at 550° C, 300 Knoop at 1200° C

MODULUS OF ELASTICITY: 393,000 MPa at 20° C, 324,000 MPa at 2050° C

FLEXURAL STRENGTH: 275 MPa at 20° C, 240 MPa at 1200° C

SHEAR MODULUS: 193,000 MPa

POISSON'S RATIO: 0.18

ELECTRICAL RESISTIVITY: 0.0001 Ohm.cm

SOLUBILITY IN WATER: none

TOXICITY: If it is ingested there is danger of liver damage. In addition, hafnium carbide will react with water or steam to form flammable vapors.

STRUCTURE: face centered cubic

APPLICATIONS: control rods for nuclear reactors

HAFNIUM OXIDE, HAFNIA, HfO_2

MELTING TEMPERATURE: 2758° C

BOILING TEMPERATURE: about 5400° C

SPECIFIC GRAVITY: 9.7

SPECIFIC HEAT: 0.32 kJ/kgK at 20° C, 0.45 kJ/kgK at 1100° C, 0.50 kJ/kgK at 1550° C

THERMAL CONDUCTIVITY: 4.3 W/mK at 20° C, 2.2 W/mK at 550° C for monoclinic; 1.6 W/mK over 20° to 800° degrees C for stabilized cubic

COEFFICIENT OF LINEAR EXPANSION: 2.9×10^{-6}/K over 22° to 1000° C, 3.7×10^{-6}/K over 1000° to 2200° C for the stabilized cubic structure

THERMAL SHOCK: It has poor resistance except when hafnia is partially stabilized in the cubic structure. Then it has good resistance

HARDNESS: 780 Knoop

TENSILE STRENGTH: 90 MPa

SHEAR STRENGTH: 110 MPa

COMPRESSIVE STRENGTH: 1,390 MPa

ELECTRICAL RESISTIVITY: 108 Ohm.cm

DIELECTRIC CONSTANT: 12 at 1 MHz

DISSIPATION FACTOR: 0.12 at 1 MHz

MAGNETIC SUSCEPTIBILITY: -2.89×10^{-7} m^3/kg

COLOR: white

SOLUBILITY IN WATER: none

TOXICITY: Liver damage is expected if hafnium oxide is ingested.

STRUCTURE: Low temperature hafnium oxide is monoclinic and transforms to tetragonal at about 1650° C. If hafnium oxide is stabilized by additives, then it is cubic and will remain so during heating or cooling.

ADDITIVES: About 6% yttrium oxide will produce a partially stabilized cubic structure in hafnium oxide. About 10% yttrium oxide is needed for it to be fully stabilized.

APPLICATIONS: It is used as an additive to other refractory materials. As an additive, it suppresses thermal expansion and phase changes.

HELIUM, He

MELTING TEMPERATURE: 0.95 K with pressure, Without pressure helium never solidifies.
BOILING TEMPERATURE: 4.2 K
SPECIFIC GRAVITY: 0.0018
SPECIFIC HEAT: 5.24 KJ/KgK
THERMAL CONDUCTIVITY: 0.0354 W/mK at 33 K, 0.0684 W/mK 87 K, 0.115 W/mK at 200 K, 0.136 W/mK at 255 K, 0.158 W/mK at 49° C
DIELECTRIC CONSTANT: 1.055 at 2 K
MAGNETIC SUSCEPTIBILITY: -5.9×10^{-9} m^3/kg
COLOR: colorless gas
CHEMICAL STABILITY: inert
TOXICITY: Like all of the gases in this family, it can interfere with the absorption of oxygen in the lungs if it occurs in sufficient quantity.
STRUCTURE: Helium is close-packed hexagonal up to 1.15 K with a pressure of 6.69 MPa. It is face centered cubic up to 16 K with a pressure of 127 MPa. It can also be body centered cubic up to 1.73 K with a pressure of 2.94 MPa.
APPLICATIONS: Helium is used for airborne balloons both as toys and for military/commercial purposes. It is also used as a food additive to fluff products.

HYDROGEN DIOXIDE, ICE, WATER, H_2O

MELTING TEMPERATURE: 0° C
BOILING TEMPERATURE: 100° C
SPECIFIC GRAVITY: 1.0 at 20° C; 0.96 at 0° C
SPECIFIC HEAT: 0.47 kJ/kgK at 50 K, 1.16 kJ/kgK at 150 K, 1.86 kJ/kgK at 250 K
THERMAL CONDUCTIVITY: 30 W/mK at 35 K, 35 W/mK at 40 K, 7w/mK at 100 K, 2.8 W/mK at 200 K
COEFFICIENT OF LINEAR EXPANSION: 7×10^{-6}/K at 100 K, 35×10^{-6}/K at 200 K, 50×10^{-6}/K at 270 K
MODULUS OF ELASTICITY: 8,900-9,900 MPa at 268 K
SHEAR MODULUS: 3,400-3,800 MPa at 268 K

168

POISSON'S RATIO: 0.31-0.36 at 268 K

DIELECTRIC CONSTANT: 100 at 268 K, 78.5 at 25° C, 34.5 at 200° C

PIEZOELECTRIC: at low temperature

FERROELECTRIC: at low temperature

REFRACTIVE INDEX: 1.33 at 20° C; 1.32 at 100° C

COLOR: clear

ACOUSTIC LONGITUDINAL WAVE VELOCITY: 3.99 km/s for ice

ACOUSTIC SHEAR WAVE VELOCITY: 1.98 km/s for ice

ACOUSTIC LONGITUDINAL WAVE IMPEDANCE: 3.66 kg/m^{2}s

TOXICITY: none

STRUCTURE: Hydrogen dioxide is hexagonal. It is held together by hydrogen bonds and van der Waals bonds which make it very sensitive to orientation for its properties.

APPLICATIONS: It finds extensive use as a building material, solvent and heat exchange medium. Its pervasive presence in our environment has actually caused it to be overlooked as a possible ceramic in more exotic applications.

IODINE, I

MELTING TEMPERATURE: 113.7^o C

BOILING TEMPERATURE: 183° C

SPECIFIC GRAVITY: 4.94

SPECIFIC HEAT: 0.22 KJ/KgK

THERMAL CONDUCTIVITY: 0.44 W/mK

COEFFICIENT OF LINEAR THERMAL EXPANSION: 93×10^{-6}/K

ELECTRICAL RESISTIVITY: 1.37×10^7 Ohm.m

DIELECTRIC CONSTANT: 11 at 140° C, 13 at 168° C, 6.8 at 400° C

MAGNETIC SUSCEPTIBILITY: -4.4×10^{-9} m^3/kg

COLOR: violet to black

CHEMICAL STABILITY: highly reactive

TOXICITY: Iodine is highly reactive and thus highly irritating to the human body. It is especially dangerous in the vapor form due to its increased ease of penetrating the body.

STRUCTURE: orthorhombic

APPLICATIONS: Iodine is used in pharmaceuticals, dyes and photography.

IRON DISULFIDE, MARCASITE, PYRITE, FeS$_2$

MELTING TEMPERATURE: 1171° C

SPECIFIC GRAVITY: 4.9 for marcasite, 5.0 for pyrite

SPECIFIC HEAT: 0.5 kJ/kgK

HARDNESS: 6.5 Mohs

MAGNETIC SUSCEPTIBILITY: $1.25X10^{-5}$ m^3/kg
COLOR: opaque yellow
SOLUBILITY IN WATER: 0.0005 g/l00ml
TOXICITY: It is a skin irritant and a suspected carcinogen if it is inhaled or ingested. Care should be taken if it is used in the finely powdered form.
STRUCTURE: Marcasite is rhombic in crystalline structure and the normal low temperature form of iron disulfide. It transforms to pyrite which is cubic at 450° C
APPLICATIONS: It is used as a source of iron, as a pigment and in magnetic devices.

KRYPTON, Kr

MELTING TEMPERATURE: 116 K
BOILING TEMPERATURE: 121 K
SPECIFIC GRAVITY: 2.8 for the solid, 2.4 for the liquid, 0.0037 for the gas
SPECIFIC HEAT: 0.248 kJ/kgK
THERMAL CONDUCTIVITY: 0.00831 W/mK at 255 K, 0.00987 at 38° C
MAGNETIC SUSCEPTIBILITY: $-4.32X10^{-9}$ m^3/kg
COLOR: colorless gas
CHEMICAL STABILITY: Krypton is inert except with fluorine.
TOXICITY: Krypton is an asphyxiant in high concentration.
STRUCTURE: face centered cubic
APPLICATIONS: Krypton is used in flourescent lights and flash bulbs.

LEAD OXIDE, LITHARGE, MASSICOT, PbO

MELTING TEMPERATURE: 890° C
MINIMUM MELTING POINT FOR BINARY SYSTEM: 865° C for alumina; 720° C for silica
SPECIFIC GRAVITY: 8.0 for massicot, 9.5 for litharge
SPECIFIC HEAT: 0.20 kJ/kgK
HARDNESS: 2 Mohs
MAGNETIC SUSCEPTIBILITY: $-5.28X10^{-7}$ m^3/kg
REFRACTIVE INDEX: 2.5-2.7
COLOR: yellow for massicot and red for litharge
SOLUBILITY IN WATER: 0.0017 g/100ml for litharge, 0.0023 g/100ml for massicot
TOXICITY: It may have the toxicity of lead but this has not been documented.
STRUCTURE: orthorhombic for massicot, tetragonal for litharge APPLICATIONS: It is used as an additive to glass to increase the index of refraction. Lead oxide also makes the glass softer and easier to cut.

LITHIUM ALUMINOSILICATE, SPODUMENE, $Li_2O \cdot Al_2O_3 \cdot 4SiO_2$

MELTING TEMPERATURE: 1421° C
SPECIFIC GRAVITY: 3.1 for alpha and 2.4 for beta
SPECIFIC HEAT: 0.44 kJ/kgK
THERMAL CONDUCTIVITY: 5.02 W/mk
COEFFICIENT OF LINEAR EXPANSION: $2X10^{-6}$/K
THERMAL SHOCK RESISTANCE: good
HARDNESS: 6.5-7 Mohs
TENSILE STRENGTH: 30 MPa
SHEAR STRENGTH: 55 MPa
COMPRESSIVE STRENGTH: 900 MPa
ELECTRICAL RESISTIVITY: 10^{10} Ohm.cm
DIELECTRIC CONSTANT: 6.4 at 1 MHz
DISSIPATION FACTOR: 0.03 at 1 MHz
REFRACTIVE INDEX: 1.6
STRUCTURE: It has a monoclinic structure for both the alpha and beta form.
The alpha form transforms into the beta form at about 700° C.
APPLICATIONS: It is used as a flux in porcelain as well as an additive to glass to reduce thermal shock.

LITHIUM FLUORIDE, LiF

MELTING TEMPERATURE: 845° C
BOILING TEMPERATURE: 1676° C
SPECIFIC GRAVITY: 2.6
SPECIFIC HEAT: 1.55 kJ/kgK
THERMAL CONDUCTIVITY: 50 W/mK at 2 K, 1,000 tq/mk at 20 K, 100
W/mK at 100 K, 12 W/mK at 20° C
COEFFICIENT OF LINEAR EXPANSION: $40X10^{-6}$/K
HARDNESS: 120 Knoop
MODULUS OF ELASTICITY: 62,000 MPa
DIELECTRIC CONSTANT: 9
MAGNETIC SUSCEPTIBILITY: $-1.33X10^{-7}$ m^3/kg
REFRACTIVE INDEX: 1.4
TRANSMISSION OF LIGHT: 10% or more through a thickness of 2 mm in the region 0.12-8.5 microns
COLOR: white
SOLUBILITY IN WATER: 0.27 g/100ml
TOXICITY: It is very nasty stuff when in solution or vaporized. The fluoride component attacks bone as well as tissue causing severe long term deterioration. The lithium is not a whole lot better behaved in the body. It primarily attacks the kidneys. Long term exposure can lead to death.

STRUCTURE: cubic
APPLICATIONS: Its most prevalent use is as a flux in porcelain glazes. It is also used as a window for sighting devices such as in space where water or water vapor are not present.

MAGNESIUM ALUMINATE, SPINEL, $MgOAl_2O_3$

MAXIMUM CONTINUOUS SERVICE TEMPERATURE: 1100^0 C; 1950^0 C in an oxidizing atmosphere

MELTING TEMPERATURE: 2135^0 C
SPECIFIC GRAVITY: 3.55

SPECIFIC HEAT: 0.84 kJ/kgK at 20^0 C, 0.88 kJ/kgK at 1040^0 C

THERMAL CONDUCTIVITY: 1.7 W/mK at 20^0 C, 17 W/mK at 200^0 C, 5 W/mK at 1300^0 C

COEFFICIENT OF LINEAR EXPANSION: $8.1X10^{-6}$/K for 20^0 to 400^0 C
THERMAL SHOCK: moderate resistance, slightly less than alumina
HARDNESS: 7.5-8 Mohs
MODULUS OF ELASTICITY: 157,000 MPa
TENSILE STRENGTH: 95 MPa
FLEXURAL STRENGTH: 145 MPa
SHEAR STRENGTH: 103 MPa
SHEAR MODULUS: 110,000 MPa
POISSON'S RATIO: 0.31
COMPRESSIVE STRENGTH: 1,710 MPa

CREEP RATE: $2.6X10^{-4}$ /hr at 1300^0 C with a load of 12 MPa

ELECTRICAL RESISTIVITY: 10^{15} Ohm.cm at 20^0 C, 10^{14} Ohm.cm at 300^0 C, 10^{11} Ohm.cm at 500^0 C

DIELECTRIC STRENGTH: 10 kv/mm
DIELECTRIC CONSTANT: 8 at 1 MHz
DISSIPATION FACTOR: 0.0003 at 1 MHz
REFRACTIVE INDEX: 1.715-1.725
COLOR: clear to white
EMITTANCE: 0.9 for 7 to 11 microns
SOLUBILITY IN WATER: none
CHEMICAL STABILITY; RESISTANCE TO:
    GLASSES: highly resistant
TOXICITY: see alumina
STRUCTURE: Cubic
APPLICATIONS: It finds usage as a refractory brick material in the glass and slag industry.

MAGNESIUM ALUMINOSILICATE, CORDIERITE, $2MgO \cdot 2Al_2O_3 \cdot 5SiO_2$

MAXIMUM CONTINUOUS SERVICE TEMPERATURE: 1200° C

MELTING TEMPERATURE: 1250° C
SPECIFIC GRAVITY: 2.1
SPECIFIC HEAT: 0.8 kJ/kgK

THERMAL CONDUCTIVITY: 3 W/mK at 20° C, 2 W/mK at 500° C

COEFFICIENT OF LINEAR EXPANSION: 2.6×10^{-6}/K for 25° to 700° C
THERMAL SHOCK: excellent resistance
HARDNESS: 6-7.5 Mohs
MODULUS OF ELASTICITY: 60,000 MPa
WEIBULL MODULUS: 10

TENSILE STRENGTH: 100 MPa for 20° to 1200° C
RUPTURE MODULUS: 105 MPa
FLEXURAL STRENGTH: 110 MPa
SHEAR STRENGTH: 117 MPa
POISSON'S RATIO: 0.2
COMPRESSIVE STRENGTH: 275 MPa

ELECTRICAL RESISTIVITY: 10^{14} Ohm.cm at 20° C, 10^5 Ohm.cm at 900° C
DIELECTRIC STRENGTH: 5 kV/mm
DIELECTRIC CONSTANT: 4.9 at 1 MHz
DISSIPATION FACTOR: 0.01 at 1 MHz
REFRACTIVE INDEX: 1.5
COLOR: gray
TOXICITY: It is essentially that of alumina.
STRUCTURE: a layered arrangement of crystalline components of magnesia, alumina and silica
APPLICATIONS: A large volume of it is used as a binder in refractory brick making. It is not generally used in precision shapes because it shrinks unpredictably during firing.

MAGNESIUM FLUORIDE, SELLAITE, AFFUON, MgF_2

MELTING TEMPERATURE: 1261° C

BOILING TEMPERATURE: 2239° C
SPECIFIC GRAVITY: 3.14
SPECIFIC HEAT: 0.523 kJ/kgK
HARDNESS: 5 Mohs
REFRACTIVE INDEX: 1.38
COLOR: gray
OPTICAL TRANSMISSION: 90% for 3 to 5 microns with a thickness of 2.54 mm and a temperature range of 20° to 500° C
SOLUBILITY IN WATER: slight in cold water
STRUCTURE: tetrahedral
TOXICITY: Like any fluoride, magnesium fluoride is highly toxic. Inhalation

173

and/or ingestion, especially when it is chronic, can lead to many health problems and even death.

APPLICATIONS: Magnesium fluoride is most used as a microwave filter.

MAGNESIUM OXIDE, MAGNESIA, PERICLASE, MgO

MAXIMUM CONTINUOUS SERVICE TEMPERATURE: 2300° C for an oxidizing atmosphere; 1700° C for a reducing atmosphere

MELTING TEMPERATURE: 2850° C

MINIMUM MELTING POINT OF BINARY SYSTEM: 1995° C for alumina; 1855° C for beryllium oxide; 2300° C for calcium oxide; 1543° C for silica; 2100° C for thorium oxide; 1600° C for titanium oxide; 2113° C for zirconium oxide; 2340° chromium oxide

SPECIFIC GRAVITY: 3.58

SPECIFIC HEAT: 0.92 kJ/kgK at 20° C, 1.31 kJ/kgK for 1100° to 2100° C

THERMAL CONDUCTIVITY: 52 W/mK at 20° C, 4 W/mK at 1100° C

COEFFICIENT OF LINEAR EXPANSION: $3.6X10^{-6}$/K over 20° to 550° C, $4.9X10^{-6}$/K over 550° to 1000° C, $5.3X10^{-6}$/K over 1000° to 2200° C

THERMAL SHOCK: poor resistance

HARDNESS: 5.5-6 Mohs; 450 Knoop at 20° C, 100 Knoop at 1000° C

MODULUS OF ELASTICITY: 345,000 MPa at 20° C, 275,000 MPa at 1100° C

TENSILE STRENGTH: 220 MPa at 20° C, 100 MPa at 1200° to 1600° C

FLEXURAL STRENGTH: 242 MPa at 20° C, 104 MPa at 1100° C

FRACTURE TOUGHNESS: 1.09 MPa.m$^{1/2}$

SHEAR STRENGTH: 138 MPa

SHEAR MODULUS: 130,000 MPa at 20° C, 62,000 MPa at 1100° C

POISSON'S RATIO: 0.18 at 20° C, 0.36-0.47 at 1100° C

COMPRESSIVE STRENGTH: 170 MPa at 1200° C, 100 MPa at 1500° C

CREEP RATE: $3X10^{-5}$/hr at 1300° C and 12 MPa load

ELECTRICAL RESISTIVITY: $9X10^{12}$ Ohm.cm at 1000° C, $7X10^{9}$ Ohm.cm at 1600° C, $2X10^{8}$ Ohm.cm at 2100° C

DIELECTRIC STRENGTH: 8.5-11 kv/mm

DIELECTRIC CONSTANT: 5.4 at 1 MHz.

DISSIPATION FACTOR: 0.0001 at 1 MHz

MAGNETIC SUSCEPTIBILITY: $-1.28X10^{-7}$ m^3/kg

REFRACTIVE INDEX: 1.73

COLOR: clear to white

SOLUBILITY IN WATER: 0.0006 g/100ml at 20° C, 0.009 g/100ml at 30° C

DIFFUSION COEFFICIENT: $4X10^{-15}$ cm.cm/sec at 1300° C, $8X10^{-13}$

cm.cm/sec at 1700° C for oxygen through magnesium oxide; 4×10^{-11} cm.cm/sec at 1400° C, 2×10^{-10} cm.cm/sec at 1600° C for magnesium through magnesium oxide

CHEMICAL STABILITY; RESISTANCE TO:
ACIDS: readily attacked
BASES: good resistance
GASES: readily reacts with water vapor to form hydrogen. It also reacts with hydrogen chloride. It has limited reactions with sulphur and carbon containing gases.
METALS: good resistance to both solid and liquid state. TOXICITY: It has use as a dietary supplement and has no ill effects on the human body in the solid form. The vapor may cause difficulties due to its reactivity.
RESPONSE TO IRRADIATION: Transgranular cracking occurs at doses of 2×10^{20} nvt and above. The amount of damage at this dose rate was less at ambient than at 100° C.
ADDITIVES: The addition of chromium oxide or aluminum oxide produces a spinel structure within the material which reduces the thermal expansion and improves the thermal shock resistance.
STRUCTURE: cubic, rock salt structure
APPLICATIONS: This is a widely used material. It is found in refractory bricks, electrical insulators, cements, and fertilizer. It is also used as an additive for plastics and rubber.

MAGNESIUM SILICAHYDRATE, STEATITE, $4MgO \cdot 5SiO_2 \cdot H_2O$

SPECIFIC GRAVITY: 2.6
THERMAL CONDUCTIVITY: 3.34 W/mK
COEFFICIENT OF LINEAR EXPANSION: 10^{-5}/K
THERMAL SHOCK: moderate resistance
TENSILE STRENGTH: 76 MPa
FLEXURAL STRENGTH: 124 MPa
SHEAR STRENGTH: 145 MPa
COMPRESSIVE STRENGTH: 628 MPa
ELECTRICAL RESISTIVITY: 10^{17} Ohm.cm
DIELECTRIC STRENGTH: 10 kv/mm
DIELECTRIC CONSTANT: 6 at 1 MHz
DISSIPATION FACTOR: 0.03 at 1 MHz
TOXICITY: based on silica
APPLICATIONS: It is used as a additive to glass.

MAGNESIUM SILICATE, FORSTERITE, $2MgO \cdot SiO_2$

MELTING TEMPERATURE: 1910° C

SPECIFIC GRAVITY: 2.8-3.2
SPECIFIC HEAT: 0.8 kJ/kgK
THERMAL CONDUCTIVITY: 10.6 W/mK
COEFFICIENT OF LINEAR EXPANSION: 10^{-5}/K
THERMAL SHOCK: poor resistance
HARDNESS: 7 Mohs
TENSILE STRENGTH: 76 MPa
SHEAR STRENGTH: 145 MPa
COMPRESSIVE STRENGTH: 593 MPa
ELECTRICAL RESISTIVITY: 10^{17} Ohm.cm
DIELECTRIC STRENGTH: 7.9-11.9 kv/mm
DIELECTRIC CONSTANT: 5.8-6.7 at 1 MHz
DISSIPATION FACTOR: 0.006 at 1 MHz
REFRACTIVE INDEX: 1.7
COLOR: white
SOLUBILITY IN WATER: none
TOXICITY: What toxicity it displays is due to its silica content.
STRUCTURE: orthorhombic
APPLICATIONS: Magnesium silicate is used for furnace lining.

MOLYBDENUM CARBIDE, Mo_2C

MAXIMUM CONTINUOUS SERVICE TEMPERATURE: 550° C in air
MELTING TEMPERATURE: 2687° C
SPECIFIC GRAVITY: 8.9
SPECIFIC HEAT: 0.34 kJ/kgK at 20° C, 0.44 kJ/kgK at 1100° C
COEFFICIENT OF LINEAR EXPANSION: 6.6×10^{-6}/K for 22° to 1950° C
HARDNESS: 1700 Knoop
COMPRESSIVE STRENGTH: 900 MPa
ELECTRICAL RESISTIVITY: 10^{-4} Ohm.cm
COLOR: white
SOLUBILITY IN WATER: none
TOXICITY: none known
STRUCTURE: hexagonal
APPLICATIONS: Molybdenium carbide has applications in the tool industry.

NEODYMIUM OXIDE, NEODYMIA, Nd_2O_3

MELTING TEMPERATURE: 2128° C
SPECIFIC GRAVITY: 7.24
SPECIFIC HEAT: 0.33 kJ/kgK
COEFFICIENT OF LINEAR EXPANSION: 11.8×10^{-6}/K for 25° to 800° C,

14.1X10^{-6}/K for 800° to 1050° C, 17.1X10^{-6}/K for 1050° to 1320° C
HARDNESS: 380-650 Knoop
COMPRESSIVE STRENGTH: 110-130 MPa
MAGNETIC SUSCEPTIBILITY: 1.28X10^{-4} m^3/kg
COLOR: light blue
SOLUBILITY IN WATER: 0.0002 g/100ml in cold water, 0.003 g/100ml in hot water
CHEMICAL STABILITY; RESISTANCE TO:
    GASES: reacts slowly with water vapor
TOXICITY: probably toxic but no conclusive evidence
STRUCTURE: hexagonal
APPLICATIONS: Neodymium oxide is used as a glass additive to produce a violet color.

NEON, Ne

MELTING TEMPERATURE: 24 K
BOILING TEMPERATURE: 27 K
SPECIFIC GRAVITY: 0.0009
SPECIFIC HEAT: 2.08 kJ/kgK
THERMAL CONDUCTIVITY: 0.0410 W/mK at 233 K, 0.0448 W/mK at 266 K,

0.0507 W/mK at 49° C

MAGNETIC SUSCEPTIBILITY: -4.2X10^{-9} m^3/kg
COLOR: colorless gas
SOLUBILITY IN WATER: dilute
CHEMICAL STABILITY: inert
TOXICITY: Like the other nobel gases, it is an asphyxiant in high concentration.
STRUCTURE: face centered cubic
APPLICATIONS: It is used extensively in light bulbs.

NIOBIUM CARBIDE, NbC

MAXIMUM CONTINUOUS SERVICE TEMPERATURE: 3450° C in helium

MELTING TEMPERATURE: 3500° C
SPECIFIC GRAVITY: 7.6

SPECIFIC HEAT: 0.3 kJ/kgK at 20° C, 0.5 kJ/kgK at 1100° C

THERMAL CONDUCTIVITY: 14 W/mK at 20° C, 29 W/mK at 1100° C

COEFFICIENT OF LINEAR EXPANSION: 7.3X10^{-6}/K for 20° to 2200° C

HARDNESS: 2300 Knoop at 20° C, 1000 Knoop at 550° C, 480 Knoop at 1100° C

MODULUS OF ELASTICITY: 414,000 MPa at 20° C, 310,000 MPa at 2200° C
TENSILE STRENGTH: 250 MPa
RUPTURE MODULUS: 240 MPa

177

FLEXURAL STRENGTH: 275 MPa

SHEAR MODULUS: 200,000 MPa at 20° C, 138,000 MPa at 1950° C

POISSON'S RATIO: 0.22 at 20° C, 0.25 at 1950° C
COMPRESSIVE STRENGTH: 2,400 MPa

ELECTRICAL RESISTIVITY: 10^{-4} Ohm.cm over most of the solid state
COLOR: black
SOLUBILITY IN WATER: none
CHEMICAL STABILITY; RESISTANCE TO:

GASES: reacts with hydrogen above 1400° C; stable with nitrogen to 2200° C
TOXICITY: unknown but suspected of being poisonous
STRUCTURE: cubic
APPLICATIONS: It is used as a carbide in cutting tools.

NIOBIUM TRIOXIDE, Nb_2O_3

MELTING TEMPERATURE: 1780° C
SPECIFIC GR4AVITY: 5

SPECIFIC HEAT: 0.5 kJ/kgK at 20° C, 0.68 kJ/kgK at 1100° C
HARDNESS: 6.5 Mohs
MODULUS OF ELASTICITY: 159,000 MPa
COLOR: blue to black
TOXICITY: Niobium trioxide is suspected of being poisonous.

NITROGEN, N

MELTING TEMPERATURE: 63 K
BOILING TEMPERATURE: 77 K
SPECIFIC GRAVITY: 0.0013
SPECIFIC HEAT: 2.08 kJ/kgK
THERMAL CONDUCTIVITY: 0.00848 W/mK at 87 K, 0.0211 W/mK at 233 K,

0.0235 W/mK at 266 K, 0.0275 W/mK at 49° C
DIELECTRIC CONSTANT: 1.454 at 70 K

MAGNETIC SUSCEPTIBILITY: -5.4×10^{-9} m^3/kg
COLOR: colorless gas
CHEMICAL STABILITY: Nitrogen is nonreactive with most things.
TOXICITY: Because it can replace oxygen in the atmosphere, it is considered an asphyxiant in high concentrations.
STRUCTURE: Nitrogen is cubic until 35 K where it transforms into closed packed hexagonal.
APPLICATIONS: Nitrogen is used as a food additive.

OXYGEN, O

MELTING TEMPERATURE: 54 K
BOILING TEMPERATURE: 90 K
SPECIFIC GRAVITY: 0.0014
SPECIFIC HEAT: 0.91 kJ/kgK
THERMAL CONDUCTIVITY: 0.00789 W/mK at 87 K, 0.0212 W/mK at 233 K,
0.0240 W/mK at 266 K, 0.0286 W/mK at 49° C, 0.0322 W/mK at 93° C
DIELECTRIC CONSTANT: 1.51 at 80 K
MAGNETIC SUSCEPTIBILITY: -1.36×10^{-6} m/kg
COLOR: colorless gas
CHEMICAL STABILITY: Oxygen is very stable in itself, however, most chemicals react with it to form a whole array of substances.
TOXICITY: It is not toxic in its normal state. If it reacts, the resulting oxides are harmful if the other component is inherently toxic.
HAZARD: Oxygen is a moderate fire and explosion hazard when it is stored under pressure.
STRUCTURE: Oxygen is orthorhombic until 23.8 K where it transforms to rhombohedral. At 43.8 K the rhombohedral becomes cubic.
APPLICATIONS: It is essential to all animal life and also to combustion.

PHOSPHORUS, P

MELTING TEMPERATURE: 44.3° C
BOILING TEMPERATURE: 280° C
SPECIFIC GRAVITY: 1.83 for cubic and 2.20 for amorphous
SPECIFIC HEAT: 0.74 kJ/kgK
THERMAL CONDUCTIVITY: 0.235 W/mK
COEFFICIENT OF LINEAR THERMAL EXPANSION: 125×10^{-6}/K
ELECTRICAL RESISTIVITY: 10^{11} Ohm.cm
DIELECTRIC CONSTANT: 4.10 at 34° C, 40.6 at 46° C, 3.86 at 85° C
MAGNETIC SUSCEPTIBILITY: -1.1×10^{-8} m^3/kg for cubic, -8.4×10^{-9} m^3/kg for amorphous
COLOR: white for cubic, red for amorphous powder
CHEMICAL STABILITY; RESISTANCE TO:
    ACIDS: Phosphorous is resistant to most acids.
    BASES: It reacts readily with most bases.
    GASES: There is little resistance to oxygen.
    SOLVENTS: Phosphorous has no reaction with water.
TOXICITY: Phosphorus vapor is easily inhaled and distributed throughout the body. It most characteristically effects the jaw bone and liver, but any tissue can be harmed by exposure.
HAZARD: It is a fire and explosion hazard. When in crystalline form it ignites with exposure to the oxygen of the air.
STRUCTURE: cubic or amorphous

APPLICATIONS: Fertilizers are the most common use of phosphorous.

PLUTONIUM DIOXIDE, PuO_2

MELTING TEMPERATURE: 2650° C
SPECIFIC GRAVITY: 11.5
SPECIFIC HEAT: 0.24 kJ/kgK at 20° C, 0.30 at 1100° C
THERMAL CONDUCTIVITY: 9 W/mK at 20° C, 5 W/mK at 1100° C
MAGNETIC SUSCEPTIBILITY: 0.00073 cgs
COLOR: yellowish green
TOXICITY: It is extremely toxic because plutonium and plutonium compounds
concentrate in the bone. The radioactive level is always high and takes a very long
time to dissipate.
APPLICATIONS: It is used as a fuel in nuclear reactors.

PORCELAIN

MELTING TEMPERATURE: approximately 1600° C
SPECIFIC GRAVITY: 2.3-2.5
SPECIFIC HEAT: 0.25 kJ/kgK
THERMAL CONDUCTIVITY: 1.7 W/mK for 100° to 1000° C
COEFFICIENT OF LINEAR EXPANSION: $5X10^{-6}$/K over 20° to 1000° C
THERMAL SHOCK: good resistance
HARDNESS: 7-8 Mohs
MODULUS OF ELASTICITY: 103,000 MPa
TENSILE STRENGTH: 59 MPa, usually about 20% higher for glazed material
RUPTURE STRENGTH: 200 MPa for 3% porosity; 125 MPa for 10% porosity
RUPTURE MODULUS: 70-80 MPa
IMPACT STRENGTH: 15 MJ absorbed on impact
FLEXURAL STRENGTH: 70 MPa
SHEAR STRENGTH: 180 MPa
POISSON'S RATIO: 0.21
COMPRESSIVE STRENGTH: 690 MPa
ELECTRICAL RESISTIVITY: 10^{14} Ohm.cm
DIELECTRIC STRENGTH: 50 kv/mm
DIELECTRIC CONSTANT: 6
DISSIPATION FACTOR: 0.0003 at 1 MHz
COLOR: white
ACOUSTIC LONGITUDINAL WAVE VELOCITY: 5.9 km/s
ACOUSTIC LONGITUDINAL WAVE IMPEDANCE: 13.5 kg/m^2s
GAS TRANSMISSION: does not transmit gas until 1400° C
CHEMICAL STABILITY; RESISTANCE TO:
    ACIDS: very good except to hydrofluoric

BASES: very good except to sodium hydroxide
SOLVENTS: very good with organic solvents
GASES: very good
METALS: good to very good in both the liquid and solid states at moderate temperatures.
TOXICITY: based on the silica content
ADDITIVES: A silica based glaze is often used to increase strength and reduce surface roughness. Beryllium oxide is used to increase thermal conductivity and decrease thermal shock.
STRUCTURE: It is a complex material combining crystalline and amorphous natural clays. The final composition usually contains mullite and quartz dispersed throughout a continuous silicate glass.
APPLICATIONS: Its low cost gives it a wide range of applications. Primarily it is used for tableware, sanitary ware and electrical insulation.

POTASSIUM ALUMINOSILICATE, ORTHOCLASE, MICROCLIME, MICA, $K_2O \cdot Al_2O_3 \cdot 6SiO_2$

MELTING TEMPERATURE: 1140-1300° C
SPECIFIC GRAVITY: 2.5-2.6
THERMAL CONDUCTIVITY: 0.5W/mK
COEFFICIENT OF LINEAR EXPANSION: $2X10^{-6}$/K
HARDNESS: 6-6.5 Mohs
COMPRESSIVE STRENGTH: 221 MPa
COEFFICIENT OF FRICTION: 1.0 for mica on mica
ELECTRICAL RESISTIVITY: 10^{16} Ohm.cm
DIELECTRIC STRENGTH: 60 kv/mm
DIELECTRIC CONSTANT: 7 at 1 MHz
DISSIPATION FACTOR: 0.004 at 1 MHz
REFRACTIVE INDEX: 1.5
BIREFRINGENCE: 0.007 for orthoclase
COLOR: white
STRUCTURE: Microcline has a triclinic crystalline structure whereas orthoclase is monoclinic.
APPLICATIONS: sighting device windows, electrical insulation

PRASEODYMIUM OXIDE, PRASEODYMIA, Pr_2O_3

MELTING TEMPERATURE: 2212° C
SPECIFIC GRAVITY: 7.07
SPECIFIC HEAT: 0.36 kJ/kgK
HARDNESS: 370-380 Knoop
MAGNETIC SUSCEPTIBILITY: $5.02X10^{-5}$ m^3/kg at 554° C

COLOR: yellow to green
SOLUBILITY IN WATER: slight
TOXICITY: Limited experimentation shows no toxicity.
STRUCTURE: Praseodymium oxide is cubic and transforms to hexagonal at about 880° C
APPLICATIONS: The traditional use has been as a colorant in porcelain. Its magnetic properties may be of more interest in the future.

RADON, Rn

MELTING TEMPERATURE: 202 K
BOILING TEMPERATURE: 211 K
SPECIFIC GRAVITY: 0.01
SPECIFIC HEAT: 0.0938 kJ/kgK
COLOR: colorless gas
CHEMICAL STABILITY: inert
TOXICITY: Radon is a common air contaminant. It is radioactive with a very short half life making it of concern to human beings. It tends to accumulate in the lungs and cause lung cancer. Good ventilation is usually the precaution recommended.
STRUCTURE: face centered cubic
APPLICATIONS: It is used to treat the cancers it can cause.

SAMARIUM OXIDE, SAMARIA, Sm_2O_3

MELTING TEMPERATURE: 2303° C
SPECIFIC GRAVITY: 8.35
SPECIFIC HEAT: 0.34 kJ/kgK at 20° C, 0.42 kJ/kgK at 560° C, 0.45 kJ/kgK at 1500° C
COEFFICIENT OF LINEAR EXPANSION: 10.3×10^{-6}/K for 30° to 1200° C
HARDNESS: 380-480 Knoop
MODULUS OF ELASTICITY: 138,000 MPa at 20° C, 103,000 MPa at 1350° C
SHEAR MODULUS: 117,000 MPa at 20° C, 110,000 MPa at 1500° C
POISSON'S RATIO: 0.32
MAGNETIC SUSCEPTIBILITY: 2.87×10^{-5} m^3/kg at 85 K, 2.46×10^{-5} m^3/kg at 170 K, 2.50×10^{-5} m^3/kg at 20° C
COLOR: white to yellow
SOLUBILITY IN WATER: none
CHEMICAL STABILITY; RESISTANCE TO:
    GASES: reacts slowly with water vapor
TOXICITY: It is suspected of causing impairment of blood clotting
STRUCTURE: Samarium oxide is cubic to 980° C. Between 980° C and 2300°

C it is monoclinic. At 2300° C it becomes hexagonal.

APPLICATIONS: It is used in the glass industry in both luminescent glasses and infrared absorbing glasses. It is also used as a neutron absorber in nuclear experiments.

SCANDIUM OXIDE, SCANDIA, Sc_2O_3

MELTING TEMPERATURE: 2388° C

SPECIFIC GRAVITY: 3.86

SPECIFIC HEAT: 0.68 kJ/kgK

COEFFICIENT OF LINEAR EXPANSION: 8.5X10^{-6}/K for 25° to 400° C, 9.7X10^{-6}/K for 400° to 1200° C

HARDNESS: 790-910 Knoop

MODULUS OF ELASTICITY: 220,000 MPa at 20° C, 186,000 MPa at 1650° C

FLEXURAL STRENGTH: 200 MPa at 20° C, 185 MPa at 700° C, 180 MPa at 1000° C, 125 MPa at 1350° C

SHEAR MODULUS: 90,000 MPa at 20° C, 83,000 MPa at 1650° C

COLOR: white

SOLUBILITY IN WATER: none

TOXICITY: Nothing has been firmly established but it is suspected of at least mild toxicity.

STRUCTURE: cubic

APPLICATIONS: It is used as an additive in glass to increase strength.

SELENIUM,Se

MELTING TEMPERATURE: 217° C for hexagonal

BOILING TEMPERATURE: 684.8^o C

SPECIFIC GRAVITY: 4.81 for hexagonal, 4.50 for monoclinic, 4.26 for amorphous, 4.28 for vitrius

SPECIFIC HEAT: 0.331 kJ/kgK

THERMAL CONDUCTIVITY: 2.04 W/mK

COEFFICIENT OF LINEAR THERMAL EXPANSION: 37X10^{-6}/K

HARDNESS: 736 Brinell

MODULUS OF ELASTICITY: 58,000 MPa.

POISSON'S RATIO: 0.447

SOLUBILITY IN WATER: none

ELECTRICAL RESISTIVITY: 1 Ohm.cm

DIELECTRIC CONSTANT: 5.4 at 250° C

MAGNETIC SUSCEPTIBILITY: -5.04X10^{-11} m^3/kg

REFRACTIVE INDEX: 2.75 to 3.06

COLOR: gray for hexagonal, red for monoclinic and amorphous, black for vitrius

CHEMICAL STABILITY; RESISTANCE TO:
    GASES: Selenium reacts with oxygen.
    SOLVENTS: It is resistant to water.
TOXICITY: In elemental form it is non-toxic. However, it is highly toxic on contact once it has reacted with hydrogen.
STRUCTURE: The amorphous is stable until it transforms to hexagonal at 60° C.
Monoclinic is the usual crystalline structure. It transforms to hexagonal at 170° C.
APPLICATIONS: Selenium finds extensive use in copying machines and solar cells.

SILICON, Si

MELTING TEMPERATURE: 1420° C

BOILING TEMPERATURE: 2355° C
SPECIFIC GRAVITY: 2.32
SPECIFIC HEAT: 0.7 kJ/kgK

COEFFICIENT OF LINEAR THERMAL EXPANSION: 2.6×10^{-6}/K
THERMAL CONDUCTIVITY: 150 W/mK

HARDNESS: 1,000 kg/mm^2
MODULUS OF ELASTICITY: 155,800 MPa
SHEAR MODULUS: 64,100 MPa
COMPRESSIVE STRENGTH: 93 MPa
POISSON'S RATIO: 0.215

ELECTRICAL RESISTIVITY: 10^{-3} Ohm.m
DIELECTRIC CONSTANT: 11.8

BREAKDOWN: 6×10^7 V/m
BAND GAP: 1.1 eV

SATURATED ELECTRON VELOCITY: 10^5 m/s
CARRIER MOBILITY:
    ELECTRON: 1.5 m^2/Vs

    HOLE: 0.06 m^2/Vs
ABSORPTION EDGE: 1.4 microns
REFRACTIVE INDEX: 3.5
REFLECTIVITY: 38% at 0.2 microns
COLOR: gray in cubic form, brown in powdered form

OPTICAL TRANSMITTANCE: 52% for 3 to 7 microns at 20° C, 48% for 3

microns at 350° C, 38% for 7 microns at 350° C, 17% at 3 microns at 500° C, and

3% at 5 microns at 500° C, in all cases the thickness was 3 mm
SOLUBILITY IN WATER: none
CHEMICAL STABILITY; RESISTANCE TO:
    ACIDS: Silicon is stable in acids except for HF.
    BASES: poor
    GASES: very good

SOLVENTS: very good

TOXICITY: Not usually free, silicon reacts readily with water or steam to form silicon dioxide which is toxic.

ADDITIVES: Even a minute amount of another element changes the electrical conductivity of silicon. Microchips are usually single crystal, stress-free and pure to the part per million of contamination. Silicon is also sensitive to additives with respect to light transmission.

STRUCTURE: cubic or powder

APPLICATIONS: Silicon is an important material in microchips.

SILICON CARBIDE, SiC

MAXIMUM CONTINUOUS SERVICE TEMPERATURE: 1650^{0} C for oxidizing atmosphere and 2320^{0} C for a reducing or inert atmosphere

SUBLIMATION TEMPERATURE: about 2700^{0} C

SPECIFIC GRAVITY: 2.7 (16% porosity); 3.2 (no porosity)

SPECIFIC HEAT: 0.65 kJ/kgK at 20^{0} C, 1.3 kJ/kgK at 1100^{0} C

THERMAL CONDUCTIVITY: 42 W/mK at 25^{0} C, 29 W/mK at 1350^{0} C, 17 W/mK at 1540^{0} C; 225 W/mK for the reaction bonded form at 20^{0} C

COEFFICIENT OF LINEAR EXPANSION: $4.5X10^{-6}$-$5.9X10^{-6}$/K for porosity of 0-16% and range of 20^{0} to 2000^{0} C

THERMAL SHOCK: good resistance at high temperature

HARDNESS: 2480 Knoop; 9.5 Mohs

MODULUS OF ELASTICITY: 280,000 MPa at 16% porosity; 420,000 MPa at 0% porosity, 339,000 MPa at 1500^{0} C

TENSILE STRENGTH: 500 MPa from 20^{0} to 1400^{0} C; 3,300 Mpa at 300^{0} C, 2,000 MPa at 1400^{0} C for fiber form

RUPTURE STRENGTH: 250 MPa at 16% porosity and over 20^{0} to 1400^{0} C; 550 MPa at 0% porosity over 20^{0} to 1000^{0} C, 450 MPa at 1400^{0} C

WEIBULL MODULUS: 10

FRACTURE TOUGHNESS: 4 MPa.m$^{1/2}$

FLEXURAL STRENGTH: 390-420 MPa at 20^{0} C, 515 MPa at 1100^{0} C

SHEAR MODULUS: 172,000 MPa

POISSON'S RATIO: 0.17

COMPRESSIVE STRENGTH: 1,000 MPa at 20^{0} C, 1,700 MPa at 550^{0} C, 2,000 MPa at 860^{0} C

ABRASION RESISTANCE: good

COEFFICIENT OF FRICTION: low

WEAR RATE: 0.01-0.05 mm.,mm/kg.km on steel, 0.0006 mm.mm/kg.km on gray cast iron; in-general good wear resistance. At low temperature the less dense form is superior because it allows for the infiltration of lubricants. At high temperature lubricants are ineffective, however, the structure can break down resulting in a film of graphite on the surface. This acts as a lubricant and greatly adds to the wear resistance.

CREEP RATE: 2×10^{-5}/hr at 1470° C and a load of 275 MPa ELECTRICAL RESISTIVITY: 0.001-0.1 Ohm.cm
DIELECTRIC CONSTANT: 15-20

MAGNETIC SUSCEPTIBILITY: -1.60×10^{-7} m^3/kg
REFRACTIVE INDEX: 2.68
BIREFRINGENCE: 0.043
COLOR: black
ACOUSTIC LONGITUDINAL WAVE VELOCITY: 13.06 km/s
ACOUSTIC SHEAR WAVE VELOCITY: 7.27 km/s

ACOUSTIC LONGITUDINAL WAVE IMPEDANCE: 42.0 kg/m^2s
SOLUBILITY IN WATER: none
CHEMICAL STABILITY; RESISTANCE TO:
    ACIDS: excellent resistance except to mixtures of hydrofluoric and nitric acids
    BASES: excellent resistance

GASES: Carbon monoxide reacts with silicon carbide above 1300° C to form silicon oxide and carbon. Silicon carbide reacts readily with hydrogen above 1430° C and sulphur containing gases above 1200° C. It also reacts slightly with nitrogen, oxygen and steam above 1150° C. The reaction of silicon carbide with oxygen forms a silicon dioxide coating which inhibits further reaction of the silicon carbide and oxygen.
    METALS: silicon carbide can be used with metals but in the liquid state the metals tend to react with any free silicon.
TOXICITY: It causes acute local irritation on contact and has been assigned a minimum tolerance level.
RESPONSE TO IRRADIATION: There is damage at even low dosage which results in decreased strength and thermal conductivity due to microcracking. In extreme cases this combination results in the crumbling of the material.
EMITTANCE: 0.85
ADDITIVES: Reaction bonded silicon carbide is porous when processed. Porosity of up to 18% can be overcome by infiltrating silicon which easily wets the ceramic. The final free silicon is less than would be expected because residual free graphite is usually present in reaction bonded silicon carbide. This graphite will react with the silicon during infiltration.

STRUCTURE: It is hexagonal closed packed until it reaches about 2000° C where it transforms to face centered cubic. If it is processed by vapor deposition it displays the same structure as pyrolytic graphite.
APPLICATIONS: It is used as ignition material since large amounts of radiant energy are released when a voltage is applied. It is also used as an abrasive for grinding wheels and for engine parts. Studies are considering it for high temperature turbine blades and vanes.

SILICON DIOXIDE, SILICA, VITREOUS SILICA, FUSED SILICA, FUSED

MAXIMUM CONTINUOUS SERVICE TEMPERATURE: 1050° C for vitreous silica; 1100° C for fiber form; 540° C for fused silica

MELTING TEMPERATURE: 1723° C for silica, 1703° C for tridymite; 1610° C for quartz and crystobalite

BOILING POINT: 2230° C

FICTIVE TEMPERATURE: 1473° C for glass

STRAIN TEMPERATURE: 1100° C

ANNEALING TEMPERATURE: 1200° C

SOFTENING TEMPERATURE: 1650° C

MINIMUM MELTING POINT FOR BINARY SYSTEM: 1546° C for alumina; 1670° C for beryllia; 1436° C for calcium oxide; 1543° C for magnesia; 1700° C for thoria; 1540° C for titania; 1675° C for zirconia; 1370° C for barium oxide; 1377° C for cobalt oxide; 1710° C for chromium oxide; 1060° C for copper oxide; 720° C for lead oxide; 1358° C for strontium oxide; 1432° C for zinc oxide

SPECIFIC GRAVITY: 2.19 for lechatelierite; 2.65 for quartz; 2.32 for cristobalite; 2.26 for tridymite; 2.2-8.0 for fused quartz depending on processing and additives; 2.2-2.5 for fiber

SPECIFIC HEAT: 0.75 kJ/kgK at 20° C and 1.05 kJ/kgK at 1000° C for fused silica; quartz has similar values but suffers an inversion to 2 kJ/kgK at 550° C

THERMAL CONDUCTIVITY: 1.38 W/mK at 20° C and 2.3 W/mK at 800° C for fused silica; 1000 W/mK at 15 K and 100 11/mK at 40 K for quartz

COEFFICIENT OF LINEAR EXPANSION: 0.55X10^{-6}/K for fused silica and most glasses of a silica base for 20° to 1000° C; 15X10^{-6}/K over 20° to 500° C for tridymite; 2OX10^{-6}/K over 20° to 500° C for quartz; 3OX10^{-6}/K over 20° to 500° C for cristobalite

THERMAL DIFFUSIVITY: 0.008-0.009 cm.cm./sec over 240 K to 800° C range for fused silica

THERMAL SHOCK: good resistance for quartz; excellent resistance for fused silica

HARDNESS: 6-7 Mohs for crystalline forms; 6.5 Mohs for glass forms

ELONGATION: 1.5-1.8% for 70 K to 300° C

MODULUS OF ELASTICITY: 70,000 MPa at 70 K, 73,000 MPa at 20° C and 82,000 MPa at 900° C for fused silica; 76,000 MPa for quartz whisker

TENSILE STRENGTH: 49,000-59,000 MPa for fused silica; 4,100 MPa for quartz whisker

RUPTURE MODULUS: 49 MPa at 20° C and 70 MPa at 850° C for fused silica

FLEXURAL STRENGTH: 70 MPa for fused silica

SHEAR MODULUS: 30,400 MPa at 70 K, 31,000 MPa at 20° C and 33,800 MPa at 900° C for fused silica

POISSON'S RATIO: 0.16 at 70 K, 0.17 at 20° C and 0.20 at 900° C for fused silica

COMPRESSIVE STRENGTH: 1,110,000-1,170,000 MPa for fused silica

WEAR RATE: 0.12 mm.Tnm.mm/kg.km for flint against steel

CREEP RATE: 0.2 /hr and a load of 12 MPa, 0.001 /hr for a load of 0.07 MPa for quartz at 1300° C

COEFFICIENT OF FRICTION: 0.9-1.0 for glass on glass; 0.5-0.7 for glass on metal

ELECTRICAL RESISTIVITY: 10^{18} Ohm.cm at 20° C, 10^{13} Ohm.cm at 200° C, 7×10^9 Ohm.cm at 400° C for volume of fused silica; 10^{19} Ohm at 25° C and 1 Ohm at 1200° C for the surface of glass with a fairly linear change with temperature

DIRECT CURRENT RESISTIVITY: 10^{15} at 20° C, 10^9 at 500° C for fused silica

DIELECTRIC STRENGTH: 35 kv/mm for fused silica

DIELECTRIC CONSTANT: 3.8 over a 20° to 200° C range for all frequencies, it maintains this value to 500° C at frequencies above 10,000 Hz for fused silica

DISSIPATION FACTOR: 0.003 over a 20° to 200° C range at 100 Hz., 0.0002 over a 20° to 200° C range at 1,000 Hz and 0.0005 over a 20° to 200° C range at 10,000 Hz for fused silica

MAGNETIC SUSCEPTIBILITY: -3.72×10^{-7} m^3/kg

REFRACTIVE INDEX: 1.49 for tridymite; 1.47 for cristobalite; 1.55 for quartz; 1.46 for lechatelierite and fused silica

TRANSMISSION OF LIGHT: about 90% transmission between wavelengths of 175-1000 nm for fused silica; 10% or more in the region from .16-4 microns through a 2 mm thickness for fused silica; 10% or more transmission in the region of 0.18-4.2 microns through a 2 mm thickness for fused quartz

BIREFRINGENCE: 3.45 for fused silica; 0.009 for quartz

COLOR: clear to cloudy white

ACOUSTIC LONGITUDINAL WAVE VELOCITY: 5.75 km/s for quartz, 5.9 km/s for glass

ACOUSTIC SHEAR WAVE VELOCITY: 2.2 km/s for quartz

ACOUSTIC LONGITUDINAL WAVE IMPEDANCE: 15.3 kg/m^{2}s for quartz, 13.0 kg/m^{2}s for glass

SOLUBILITY IN WATER: none

GAS TRANSMISSION RATE: 7×10^{-9} cm.cm.cm/sec/cm.cm/mm thickness/cm mercury pressure difference (or unit) for helium through glass at 20° C; 3.8×10^{-11} unit at 200° C, 5.3×10^{-10} unit at 400° C, 1.25×10^{-9} unit at 600° C and 2.1×10^{-9} unit at 700° C for hydrogen through fused silica; 2.8×10^{-10} unit at 600° C and 4.2×10^{-10} unit at 700° C for neon through fused silica; argon and nitrogen do not flow through fused silica at detectable rates

DIFFUSION COEFFICIENT: $8X10^{-15}$ cm.cm/sec at 830° C, $4X10^{-13}$ cm.cm./sec at 1400° C for oxygen through silica

ULTRASONIC PROPERTIES:

SHEAR WAVE VELOCITY: 376,000 cm/sec for fused silica

COMPRESSIVE WAVE VELOCITY: 595,000 cm/sec for fused silica

SHEAR WAVE IMPEDANCE: 827,000 gm/cm.cm/sec for fused silica

COMPRESSIVE WAVE IMPEDANCE: 1,309,000 gm/cm.cm/sec for fused silica

SONIC ATTENUATION: 1 db/cm MHz max

INTERNAL DAMPING: 10,000

CHEMICAL STABILITY; RESISTANCE TO:

ACIDS: good resistance with the exception of hydrofluoric acid to all forms. Most glasses are subject to oxides being leached from the surface by acids but fused silica is resistant to this.

BASES: much more reactive with bases than acids although attack is not appreciable unless the base is concentrated

SOLVENTS: water tends to devitrify glass by dissolving the oxide components

OILS AND FATS: excellent resistance

GASES: hydrogen reacts above 1100° C to form volatile products

METALS: titanium slightly bonds at ambient to fused silica; silica embrittles metals at high temperature; has poor resistance to metals in liquid state

TOXICITY: Free silica in the form of dust is very harmful if inhaled. The length of time for silicosis to develop depends on the individual and the working conditions. Sand-blasting can cause major lung damage within as little as two years of exposure. There seems to be no danger in ingesting silica and it is an approved food additive for animals.

RESPONSE TO IRRADIATION: It is very effective as a radiation resistant material and does not turn brown as most other glasses do under high doses.

EMISSIVITY: 0.81% at 253° C, 0.72% at 451° C, 0.56% at 693° C for 6 mm thick vitreous silica

ADDITIVES: Many, many ceramics are added for processing, coloration or structural modification to silicon dioxide. A few are given here as examples. Boron oxide reduces the viscosity of the glassy state. The addition of 7.5% titania will reduce the coefficient of thermal expansion of glass to almost zero. Copper is added in colloidal dimensions to glass such that the color centers produce a ruby red. Silver in suspension produces yellow or amber in glass. Gold produces a particularly rich red in glass; the amounts needed are on the order of 0.05%. These additives also serve as nucleation sites so care must be taken to keep the glass highly viscous if devitrification is not desired. The addition of either lithium oxide or magnesium oxide will encourage crystallization of the glass. Aluminum oxide or zinc oxide will inhibit crystalline formation.

STRUCTURE: Silica forms connected tetrahedra which are repeated in both the crystalline and glassy states. Cristobalite is a crystalline mineral which is tetragonal below 220° C and cubic above it. Tridymite is crystalline and hexagonal at all temperatures but changes its unit cell size at 250° C. Quartz is trigonal to 573° C where it changes to hexagonal. Thus, all three crystalline forms of silicon dioxide,

189

cristobalite, tridymite and quartz, exist in a low and a high temperature structure. The high temperature forms are more symmetric but less dense. The transformation from low to high temperature structure is rapid and reversible. However, a transformation between phase types is slow and sometimes almost impossible to complete.

Lechatelierite is a glassy state mineral. Fused silica is vitreous silicon dioxide in the fully dense glass form. It is made by vapor deposition of silicon tetrachloride reacted with oxygen above 1500° C The result is a translucent glass which begins to crystallize above 100° C Fused quartz is the vitreous silicon dioxide product of melting quartz or silicon bearing sand and not allowing it to crystallize. The result is a transparent glass.

APPLICATIONS: The extreme purity of fused silica plus its ease of formation into fibers allows for extensive fiber optics uses in communications networks. It is also used in furnace linings and is the principal ingredient in glass and glazes. Silica is not generally used in the crystalline phase because of the volume changes associated with each phase change. Quartz is the only extensively used phase and it is used with a high porosity to accommodate the phase change. Quartz has found high temperature applications as a container for various substances especially in laboratory environments. High purity silica glass-ceramics are used for medical implants.

SILICON NITRIDE, HOT PRESSED(HPSN), SINTERED(SSN), REACTION BONDED(RBSN), Si_3N_4

MAXIMUM CONTINUOUS SERVICE TEMPERATURE: 1200° C for oxidizing atmosphere; 1870° C for a reducing or inert atmosphere

MELTING TEMPERATURE: 1900° C, it also decomposes

SPECIFIC GRAVITY: 2.2-2.5 for RBSN; 3.18 for HPSN; 3.3 for SSN SPECIFIC HEAT: 0.6 kJ/kgK at 20° C, 1.2 kJ/kgK at 1100° C

THERMAL CONDUCTIVITY: 10-15 W/mK at 20° C, 5.0 W/mK at 1350° C for RBSN; 25 W/mK at 20° C, 15 W/mK at 1300° C for HPSN; 33 W/mK at 20° C, 18 W/mK at 600° C for SSN

COEFFICIENT OF LINEAR EXPANSION: $3X10^{-6}$/K for 0-40% porosity over 0° to 1400° C; $0.6X10^{-6}$/K for 70-300° K

THERMAL SHOCK: very resistant for RBSN; excellent for HPSN HARDNESS: 1954 Knoop for HPSN; 2900 Knoop for RBSN; 9.5 Mohs MODULUS OF ELASTICITY: 320,000 MPa at 0% porosity; 120,000 MPa at 40% porosity; 310,000 MPa for HPSN; 170,000 MPa for RBSN for 20° to 1100° C

TENSILE STRENGTH: 375 MPa at 20° C, 150 MPa at 1400° C for HPSN; 200 MPa for 20° to 1400° C; 270 MPa from 20° to 1400° C for RBSN

RUPTURE STRENGTH: 800 MPa at 20° C, 725 MPa at 600° C for SSN; 830 MPa at 20° C, 805 MPa at 600° C for TIPSN

RUPTURE MODULUS: 700 MPa at 20° C, 650 MPa at 1000° C, 400 MPa at 1400° C and 0% porosity; 150 MPa at 20° to 1400° C and 40% porosity; 190-215 MPa for RBSN; 690 MPa for HPSN

FLEXURAL STRENGTH: 989 MPa at 20° C, 435 MPa at 1370° C; 700 MPa for HPSN; 235 MPa for RBSN

SHEAR STRENGTH: 697 MPa

SHEAR MODULUS: 62,000 MPa for 20° to 1100° C

WEIBULL MODULUS: 23; 12 for RBSN; 15 for HPSN

WEIBULL NUMBER: 250 MPa for a failure stress range of 200 to 300 MPa

FRACTURE TOUGHNESS: 4 MPa.m$^{1/2}$

POISSON'S RATIO: 0.22 at 20° C, 0.14 at 1100° C

COMPRESSIVE STRENGTH: 500-620 MPa

WEAR RESISTANCE: very good

CREEP RATE: 10^{-3} at 1350° C for 6.9 MPa load, 10^{-3} at 1250° C for 52 MPa load

ELECTRICAL RESISTIVITY: 10^{14} Ohm.cm at 20° C, 10^{9} Ohm.cm at 300° C, 10^{6} Ohm.cm at 500° C, 10^{3} Ohm.cm at 900° C

DIELECTRIC STRENGTH: 17 kv/mm

DIELECTRIC CONSTANT: 6.1 at 1 MHz

DISSIPATION FACTOR: 6×10^{-4} at 1 MHz

COLOR: gray

ACOUSTIC LONGITUDINAL WAVE VELOCITY: 11.0 km/s

ACOUSTIC SHEAR WAVE VELOCITY: 6.25 km/s

ACOUSTIC LONGITUDINAL WAVE IMPEDANCE: 36.0 kg/m^{2}s

SOLUBILITY IN WATER: none

CHEMICAL STABILITY; RESISTANCE TO:

ACIDS: resistant except to hydrofluoric acid

BASES: very good

GASES: Silicon nitrate is slightly reactive to air above 1000° C. It also reacts with the vaporized phase of graphite above this temperature

METALS: It is not stable in the presence of liquid magnesium, copper, iron, cobalt, vanadium, manganese and platinum.

TOXICITY: unknown

EMITTANCE: 0.9

STRUCTURE: It is usually an amorphous gray powder which is pressed into the desired shape and heat treated. It can also be processed by vapor deposition to produce a structure similar to pyrolytic graphite. When crystalline, the alpha phase is hexagonal and converts to the beta phase, which is also hexagonal, at 1400° C

APPLICATIONS: It does not shrink on firing so it can be precision casted to the desired final dimensions. This makes it a useful material for gas turbine, air purifier and diesel engine parts.

SILVER BROMIDE, BROMYRITE, AgBr

MELTING TEMPERATURE: 432° C

DECOMPOSITION TEMPERATURE: 1300° C
SPECIFIC GRAVITY: 6.47
SPECIFIC HEAT: 0.292 kJ/kgK
THERMAL CONDUCTIVITY: 1.21 W/mK

COEFFICIENT OF LINEAR THERMAL EXPANSION: $30X10^{-6}$/K
HARDNESS: 7 Knoop
TENSILE YIELD STRENGTH: 26.2 MPa
MODULUS OF ELASTICITY: 31,970 MPa
BULK MODULUS: 44,030 MPa
DIELECTRIC CONSTANT: 13.1
REFRACTIVE INDEX: 2.42 at 0.4 microns, 2.31 at 0.5 microns, 2.25 at 0.6
microns, 2.17 at 10 microns, 2.16 at 13 microns
OPTICAL TRANSMISSION RANGE: 0.45 to 35 microns
COLOR: pale yellow

SOLUBILITY IN WATER: $8.4X10^{-6}$ g/cc
TOXICITY: Silver bromide is highly irritating on contact.
HAZARD: If exposed to fire it can produce toxic fumes.
STRUCTURE: face centered cubic
APPLICATIONS: Silver bromide is photosensitive and finds many uses for this
attribute.

SILVER CHLORIDE, CERARGYRITE, AgCl

MELTING TEMPERATURE: 455° C
BOILING TEMPERATURE: 1550° C
SPECIFIC GRAVITY: 5.56
SPECIFIC HEAT: 0.355 kJ/kgK
THERMAL CONDUCTIVITY: 1.15 W/mK

COEFFICIENT OF LINEAR THERMAL EXPANSION: $31X10^{-6}$/K
HARDNESS: 9.5 Knoop
TENSILE YIELD STRENGTH: 26.2 MPa
MODULUS OF ELASTICITY: 19,980 MPa
SHEAR MODULUS: 7,099 MPa
BULK MODULUS: 44,040 MPa
POISSION'S RATIO: 0.4
REFRACTIVE INDEX: 2.07 at 0.6 microns, 1.98 at 10 microns
OPTICAL TRANSMISSION RANGE: 0.4 to 25 microns
COLOR: white
SOLUBILITY IN WATER: cold water 0.0001 g/100ml, hot water 0.002 g/100ml.
TOXICITY: Silver chloride is highly irritating on contact.
HAZARD: If exposed to fire it can produce toxic fumes.
STRUCTURE: face centered cubic
APPLICATIONS: Silver chloride is photosensitive and finds many uses for this

attribute.

SILVER IODIDE, IODYRITE, AgI

MELTING TEMPERATURE: 552° C
BOILING TEMPERATURE: 1506° C
SPECIFIC GRAVITY: 5.67
REFRACTIVE INDEX: 2.22
COLOR: yellow
SOLUBILITY IN WATER: slight
TOXICITY: Silver iodide is highly irritating on contact.
HAZARD: If exposed to fire it can produce toxic fumes.
STRUCTURE: The alpha phase of silver iodide is hexagonal. It converts to the beta phase, which is cubic, at 146° C.
APPLICATIONS: Silver iodide is photosensitive and finds many uses for this attribute.

SILVER NITRATE, $AgNO_3$

MELTING TEMPERATURE: 212° C
BOILING TEMPERATURE: 444° C
SPECIFIC GRAVITY: 4.35
TOXICITY: Silver nitrate is highly irritating on contact.
HAZARD: If exposed to fire it can produce toxic fumes. If exposed to ammonia, powdered metals, or organics it can explode.
STRUCTURE: rhombic
APPLICATIONS: Silver nitrate is photosensitive and finds many uses for this attribute. It also has been used as a medication.

SODIUM CHLORIDE, HALITE, NaCl

MELTING .TEMPERATURE: 801° C
BOILING TEMPERATURE: 1413° C
SPECIFIC GRAVITY: 2.17
REFRACTION INDEX: 1.5
COLOR: clear
SOLUBILITY IN WATER: cold water 36 g/100ml, hot water 40 g/100ml.
ACOUSTIC LONGITUDINAL WAVE VELOCITY: 4.78 km/s
ACOUSTIC LONGITUDINAL WAVE IMPEDANCE: 10.37 kg/m^{2}s
TOXICITY: none
STRUCTURE: cubic
APPLICATIONS: Salt is not only essential to human life, it is the standard for

many scientific measurements.

STRONTIUM OXIDE, SrO

MELTING TEMPERATURE: 2430° C

BOILING TEMPERATURE: about 3000° C

SPECIFIC GRAVITY: 4.7

SPECIFIC HEAT: 0.41 kJ/kgK at 20° C, 0.53 kJ/kgK at 1100° C

THERMAL CONDUCTIVITY: 5 W/mK at 20° C, 2 W/mK at 1100° C

COEFFICIENT OF LINEAR EXPANSION: 13.5X10^{-6}/K from 20° to 650° C,
14.4X10^{-6}/K from 650° to 1200° C

HARDNESS: 3.5 Mohs

MODULUS OF ELASTICITY: 120,000 MPa

MAGNETIC SUSCEPTIBILITY: -4.40X10^{-7} m^3/kg

REFRACTIVE INDEX: 1.81

COLOR: white to gray

SOLUBILITY IN WATER: 0.69 g/100ml in cold water; 22.85 g/100ml in hot
water

CHEMICAL STABILITY; RESISTANCE TO:

   SOLVENTS: reacts readily with water

   GASES: on heating it reacts with air, carbon dioxide and sulphur dioxide

TOXICITY: It is moderately caustic to the skin but otherwise not toxic.

STRUCTURE: cubic

APPLICATIONS: It is used as a colorant in glass.

SULFUR,S

MELTING TEMPERATURE: 95.5^o for orthorhombic, 112.8^o C for monoclinic

BOILING TEMPERATURE: 444.674^o C

SPECIFIC GRAVITY: 2.07

SPECIFIC HEAT: 0.706 kJ/kgK

THERMAL CONDUCTIVITY: 0.269 W/mK

ELECTRICAL RESISTIVITY: 2X10^{13} Ohm.cm

DIELECTRIC CONSTANT: 3.52 at 118° C, 3.48 at 231° C

MAGNETIC SUSCEPTIBILITY: -7.67X10^{-11} m^3/kg for orthrorhombic,
-7.35X10^{-11} m^3/kg for monoclinic

REFRACTIVE INDEX: 1.00

COLOR: yellow

ACOUSTIC LONGITUDINAL WAVE VELOCITY: 1.35 km/s

ACOUSTIC LONGITUDINAL WAVE IMPEDANCE: 2.7 kg/m^{2}s

SOLUBILITY IN WATER: none

TOXICITY: Sulfur is an irritant in dust form.

STRUCTURE: It is orthorhombic until 93.7° C where it transforms into mono-clinic. Sulfur can also be amorphous.
APPLICATIONS: It is used in batteries and the vulcanization of rubber.

TANTALUM CARBIDE, TaC

MAXIMUM CONTINUOUS SERVICE TEMPERATURE: 2200° C in nitrogen, 3750° C in helium
MELTING TEMPERATURE: 3880° C
BOILING TEMPERATURE: 5500° C
SPECIFIC GRAVITY: 13.9
SPECIFIC HEAT: 0.17 kJ/kgK at 20° C, 0.21 kJ/kgK at 1100° C
THERMAL CONDUCTIVITY: 32 W/mK
COEFFICIENT OF LINEAR EXPANSION: 6.5×10^{-6}/K for 20° to 2200° C
THERMAL SHOCK: moderate resistance
HARDNESS: 1600 Knoop at 0° C, 1500 Knoop at 400° C, 800 Knoop at 800° C
MODULUS OF ELASTICITY: 511,000 MPa at 20° C, 455,000 MPa at 1650° C, 380,000 MPa at 2200° C
TENSILE STRENGTH: 97 MPa
RUPTURE STRENGTH: 200 MPa
FLEXURAL STRENGTH: 275 MPa at 20° C, 311 MPa at 1200° C
SHEAR MODULUS: 207,000 MPa
POISSON'S RATIO: 0.24
ELECTRICAL RESISTIVITY: 10^{-4} Ohm.cm
COLOR: black
SOLUBILITY IN WATER: none
CHEMICAL STABILITY; RESISTANCE TO:
  ACIDS: slightly reactive
  BASES: very good
  GASES: attacked by hydrogen bearing gases above 1350° C
TOXICITY: none known
EMITTANCE: 0.2
STRUCTURE: face centered cubic
APPLICATIONS: It is used as a heating element in electric furnaces.

TANTALUM OXIDE, Ta_2O_5

MELTING TEMPERATURE: 1872° C
SPECIFIC GRAVITY: 8.2
SPECIFIC HEAT: 0.2 kJ/kgK at 20° C, 0.35 kJ/kgK at 1200° C
MAGNETIC SUSCEPTIBILITY: -4.02×10^{-7} m^3/kg

COLOR: clear
SOLUBILITY IN WATER: none
TOXICITY: none
STRUCTURE: orthorhombic
APPLICATIONS: It is an additive to optical glass. It is also used for its magnetic properties.

TELLURIUM, Te

MELTING TEMPERATURE: 452^0 C for rhombic, 449^0 C for amorphous

BOILING TEMPERATURE: 1390^0 C for rhombic, 990^0 C for amorphous
SPECIFIC GRAVITY: 6.25 for rhombic, 6.0 for amorphous
SPECIFIC HEAT: 0.20 kJ/kgK
THERMAL CONDUCTIVITY: 2.35 W/mK

COEFFICIENT OF LINEAR THERMAL EXPANSION: 16.8×10^{-6}/K
HARDNESS: 180 Brinell
MODULUS OF ELASTICITY: 47,100 MPa
POISSON'S RATIO: 0.16 to 0.3
ELECTRICAL RESISTIVITY: 0.44 Ohm.sm

MAGNETIC SUSCEPTIBILITY: -4.9×10^{-11} m^3/kg
SOLUBILITY IN WATER: none
REFLECTIVITY: 50%
REFRACTIVE INDEX: 1.00
COLOR: silver when rhombic, black when amorphous
TOXICITY: It has a low overall toxicity if it is ingested; however, the side effects are nasty.
FIRE HAZARD: Tellurium is a moderate hazard in dust form.
STRUCTURE: rhombic or amorphous
APPLICATIONS: Tellurium is a p-type semi-conductor that is used in xerography. It is also used in percussion caps.

THALLIUM CHLORIDE, TlCl

MELTING TEMPERATURE: 430^0 C

BOILING TEMPERATURE: 720^0 C
SPECIFIC GRAVITY: 7.00
SPECIFIC HEAT: 0.218 kJ/kgK
THERMAL CONDUCTIVITY: 0.75 W/mK

COEFFICIENT OF LINEAR THERMAL EXPANSION: 53×10^{-6}/K
HARDNESS: 12.8 Knoop
TENSILE YIELD STRENGTH: 20.7 MPa
ELASTIC MODULUS: 31,750 MPa
SHEAR MODULUS: 7,580 MPa
POISSON'S RATIO: 0.276

DIELECTRIC CONSTANT: 31.9

REFRACTIVE INDEX: 2.25 at 0.6 microns, 2.19 at 10 microns
OPTICAL TRANSMISSION RANGE: 0.5 to 30 microns
COLOR: white
SOLUBILITY IN WATER: very slight
TOXICITY: Thallium chloride is an irritant to eyes and mucous membranes. It is especially noxious if heated, and it can be fatal if ingested.
STRUCTURE: cubic
APPLICATIONS: Thallium chloride is photosensitive and finds many uses for this attribute.

THORIUM CARBIDE, ThC_2

MELTING TEMPERATURE: 2655° C

BOILING TEMPERATURE: approximately 5000° C
SPECIFIC GRAVITY: 8.96
SPECIFIC HEAT: 0.2 kJ/kgK
THERMAL CONDUCTIVITY: 24 W/mK

COEFFICIENT OF LINEAR EXPANSION: 8.5×10^{-6}/K
HARDNESS: 800 Knoop
COLOR: yellow
SOLUBILITY IN WATER: dilute
CHEMICAL STABILITY; RESISTANCE TO:
    GASES: reacts with water vapor
TOXICITY: carcinogenic if ingested due to radioactivity

STRUCTURE: monoclinic to 1400° C were it transforms to tetragonal. From tetragonal it transforms to cubic at 1500° C.
APPLICATIONS: It is used as a nuclear fuel.

THORIUM DIOXIDE, THORIA, THORIANITE, ThO_2

MAXIMUM CONTINUOUS SERVICE TEMPERATURE: 2500° C in an oxidizing atmosphere

MELTING TEMPERATURE: 3300° C

BOILING TEMPERATURE: 4400° C

MINIMUM MELTING POINT OF BINARY SYSTEMS: 1750° C for alumina; 2150° C for beryllium oxide; 2300° C for calcium oxide; 2100° C for magnesium oxide; 1700° C for silicon dioxide; 1630° C for titanium dioxide; 2680° C for zirconium dioxide
SPECIFIC GRAVITY: 9.7-9.9

SPECIFIC HEAT: 0.23 kJ/kgK at 20° C, 0.30 kJ/kgk at 1200° C

THERMAL CONDUCTIVITY: 3.6 W/mK from 200° to 1000° C

COEFFICIENT OF LINEAR EXPANSION: 10^{-5}/K between 100° to 1000° C

THERMAL SHOCK: poor to moderate resistance

HARDNESS: 6.5 Mohs; 550 Knoop at 20° C, 200 Knoop at 550° C, 70 Knoop at 1650° C

MODULUS OF ELASTICITY: 242,000 MPa at 20° C, 207,000 MPa at 1200° C

TENSILE STRENGTH: 97 MPa

FLEXURAL STRENGTH: 104 MPa

SHEAR STRENGTH: 131 MPa

SHEAR MODULUS: 83,000 MPa at 20° C, 69,000 MPa at 1200° C

COMPRESSIVE STRENGTH: 1,380 MPa at 22° C, 1,100 MPa at 400° C, 490 MPa at 800° C., 193 MPa at 1200° C

DIFFUSION RATE: 10^{-14} cm.cm/sec at 1600° C, 10^{-13} cm.cm/sec at 1800° C for thorium; 3×10^{-15} cm.cm/sec at 1600° C, 10^{-13} cm.cm/sec at 1800° C for oxygen

CREEP RATE: 10^{-3}/hr at 1300° C for a load of 12 MPa; 10^{-4}/hr at 1960° C for a load of 35 MPa, 10^{-3}/hr at 2080° C for a load of 35 MPa, 10^{-2}/hr at 2175° C for a load of 35 MPa

ELECTRICAL RESISTIVITY: 10^{13} Ohm.cm at 20° C, 10^{12} Ohm.cm at 500° C

DIELECTRIC STRENGTH: 5 kv/mm

DIELECTRIC CONSTANT: 13.5 at 1 MHz

DISSIPATION FACTOR: 0.004 at 1 MHz

MAGNETIC SUSCEPTIBILITY: -2.01×10^{-7} m^3/kg

REFRACTIVE INDEX: 2.20 for thorianite

COLOR: white

SOLUBILITY IN WATER: none

CHEMICAL STABILITY; RESISTANCE TO:

   ACIDS: not resistant

   BASES: good resistance

   GASES: slightly reactive with hydrogen, sulphur and carbon containing gases

   METALS: excellent resistance to metals in both solid and liquid states.

Thorium oxide is especially used with titanium.

TOXICITY: thorianite is radioactive

STRUCTURE: simple cubic

APPLICATIONS: It is used as an additive in crucibles, refractories and optical glass.

THORIUM FLUORIDE, ThF_4

MELTING TEMPERATURE: 1111° C

SPECIFIC GRAVITY: 6.32

COEFFICIENT OF LINEAR EXPANSION: 10^{-6}/K

THERMAL SHOCK: extremely resistant
REFRACTIVE INDEX: 1.65
COLOR: white
TOXICITY: Thorium is radioactive, however, no other effects of the thorium have been determined. Fluoride compounds are always toxic.
STRUCTURE: cubic
APPLICATIONS: There are military applications due to its index of refraction coupled with thermal shock resistance.

TIN DIOXIDE, CASSITERITE, SnO_2

MAXIMUM CONTINUOUS SERVICE TEMPERATURE: 1400° C in an oxidizing atmosphere

MELTING TEMPERATURE: 1630° C

SUBLIMATION TEMPERATURE: 1800°-1900° C
SPECIFIC GRAVITY: 7.0
SPECIFIC HEAT: 0.35 kJ/kgK
THERMAL SHOCK: good resistance
HARDNESS: 6-7 Mohs
MAGNETIC SUSCEPTIBILITY: $-5.15X10^{-7}$ m^3/kg
REFRACTIVE INDEX: 2
COLOR: white
SOLUBILITY IN WATER: none
CHEMICAL STABILITY; RESISTANCE TO:
    ACIDS: very good
    BASES: very good
    GLASS: especially resistant to glasses and slags
    GASES: reactive with reducing gases
    METALS: poor resistance to liquid metals and graphite
TOXICITY: The dust form is mildly irritating but is not toxic.
WEAKNESSES: susceptible to failure due to mechanical shock
STRUCTURE: tetragonal, hexagonal or rhombic
APPLICATIONS: It is used in the glass industry as an electrode for electronic melting of lead glasses. It is considered the best all around opacifier for glazes. Tin dioxide is also a coloring agent in glazes. With chromates and lime it produces colors that range from pink to maroon. With vanadium compounds it produces yellows and with gold chloride it produces purple.

TITANIUM CARBIDE, TiC

MAXIMUM CONTINUOUS SERVICE TEMPERATURE: 1500° C in hydrogen

MELTING TEMPERATURE: 2940°-3250° C

BOILING TEMPERATURE: 4820° C

SPECIFIC GRAVITY: 4.92

SPECIFIC HEAT: 0.51 kJ/kgK at 20° C, 0.9 kJ/kgK at 1100° C

THERMAL CONDUCTIVITY: 20.5 W/mK at 25° C, 46.9 W/mK at 2200° C

COEFFICIENT OF LINEAR EXPANSION: 9.1×10^{-6} /K for 20° to 2000° C

HARDNESS: 3300 Knoop at 0° C, 2470 Knoop at 20° C, 1000 Knoop at 500° C, 400 Knoop at 880° C

MODULUS OF ELASTICITY: 476,000 MPa at 25° C, 352,000 MPa at 1000° C

TENSILE STRENGTH: 193 MPa at 25° C, 90 MPa at 550° C, 55 MPa at 1100° C

RUPTURE STRENGTH: 1,330 MPa
RUPTURE MODULUS: 207 MPa

FLEXURAL STRENGTH: 414 MPa at 20° C, 242 MPa at 1100° C
SHEAR STRENGTH: 1,200 MPa
SHEAR MODULUS: 186,000 MPa
POISSON'S RATIO: 0.19

COMPRESSIVE STRENGTH: 1,640 MPa at 20° C, 228 MPa at 1600° C, 90 MPa at 2200° C

CREEP RATE: 0.02/hr at 7 MPa and 2300° C; 0.2/hr at 7 MPa and 2600° C
WEAR RATE: 0.1 mm..mm/kg.km on steel

WELDING ON TEMPERATURE: 1350° C to steel

ELECTRICAL RESISTIVITY: 6×10^{-5} Ohm.cm at 20° C, 1.1×10^{-4} Ohm.cm at 800° C, 1.25×10^{-4} Ohm.cm at 1000° C

DIELECTRIC CONSTANT: 100-110 from -20° to 160° C

MAGNETIC SUSCEPTIBILITY: 1.01×10^{-7} m^3/kg
COLOR: gray
ACOUSTIC LONGITUDINAL WAVE VELOCITY: 8.27 km/s
ACOUSTIC SHEAR WAVE VELOCITY: 5.16 km/s

ACOUSTIC LONGITUDINAL WAVE IMPEDANCE: 42.6 kg/m^2s
SOLUBILITY IN WATER: none
CHEMICAL STABILITY; RESISTANCE TO:
    ACIDS: poor to moderate
    BASES: poor to moderate

    GASES: It reacts readily with air and carbon dioxide above 1200° C. The reaction is slight with carbon monoxide, steam and oxygen above 450° C. Nitrogen produces a slight reaction above 1000° C.

    METALS: extremely resistant to solid and liquid metals to high temperature
TOXICITY: Titanium carbide can be considered a nuisance dust but does not display toxicity. On decomposition, it is the carbon that is the possible hazard.
EMITTANCE: 0.5
STRUCTURE: face centered cubic
APPLICATIONS: Titanium carbide is used for gas turbine blades and pump parts.

TITANIUM DIOXIDE, TITANIA, OCTAHEDRITE, RUTILE, ANATASE, BROOKITE, TiO_2

MAXIMUM CONTINUOUS SERVICE TEMPERATURE: 1600° C in oxidizing atmosphere; 1000° C in reducing or inert atmosphere

MELTING TEMPERATURE: 1840° C

BOILING TEMPERATURE: 2500°-3000° C

MINIMUM MELTING POINT OF BINARY SYSTEMS: 1720° C for alumina; 1700° C for beryllium oxide; 1420° C for calcium oxide; 1600° C for magnesium oxide; 1540° C for silicon dioxide; 1630° C for thorium oxide; 1750° C for zirconium dioxide

SPECIFIC GRAVITY: 4.2-5.5 for rutile; 3.3-4 for anatase and octahedrite; 4.1-4.2 for brookite

SPECIFIC HEAT: 0.76 kJ/kgK at 20° C, 0.97 kJ/kgK at 1100° C

THERMAL CONDUCTIVITY: 11 W/mK at 20° C, 7 W/mK at 1100° C

COEFFICIENT OF LINEAR EXPANSION: $9.3X10^{-6}$ /K for 20° to 400° C

THERMAL SHOCK: resistance parameter is 0.2

HARDNESS: 6-6.5 Mohs for rutile; 5.5-6 Mohs for anatase and brookite; 1000 Knoop at 20° C, 200 Knoop at 1000° C

MODULUS OF ELASTICITY: 276,000 MPa

TENSILE STRENGTH: 69 MPa

FLEXURAL STRENGTH: 125 MPa; 275 MPa for bending

FRACTURE TOUGHNESS: $2.5MPa.m^{1/2}$

SHEAR MODULUS: 111,000 MPa

POISSON'S RATIO: 0.28

COMPRESSIVE STRENGTH: 690 MPa

CREEP RATE: 0.15/hr for a 25 MPa load, 1.98/hr for a 83 MPa load at 900° C; 0.37/hr for a 25MPa. load, 3.8/hr for a 33 MPa load at 950° C

ELECTRICAL RESISTIVITY: 60,000 Ohm.cm at 20° C, 11,700 Ohm.cm at 900° C, 7,500 Ohm.cm at 1000° C, 4,400 Ohm.cm at 1300° C

DIELECTRIC STRENGTH: 6 kv/mm

DIELECTRIC CONSTANT: 60

MAGNETIC SUSCEPTIBILITY: $7.41X10^{-8}$ m^3/kg

REFRACTIVE INDEX: 2.5-2.9

BIREFRINGENCE: 0.287

TRANSMISSION OF LIGHT: At least 10% of incident light is transmitted through a 2mm thickness in the region .43-6.2 microns.

COLOR: brookite is white, octahedrite and anatase are both brown

SOLUBILITY IN WATER: none

DIFFUSION COEFFICIENT: 10^{-14} cm.cm/sec at 850° C, $8X10^{-12}$ cm.cm/sec at 1350° C for oxygen through titanium dioxide

CHEMICAL STABILITY; RESISTANCE TO:
ACIDS: resistant
BASES: readily attacked
GASES: readily reacts with hydrogen
METALS: readily reacts with liquid metals
TOXICITY: a nuisance dust which is not harmful
EMITTANCE: 0.83 at 177^0 C, 0.89 at 1027^0 C
STRUCTURE: Octahedrite, rutile and anatase are tetragonal whereas brookite is orthorhombic. Anatase can exist in two forms until 642^0 C when it converts to a single form. This then changes phase at 915^0 C to form rutile. The rutile changes to brookite at 1300^0 C. These phase changes are sluggish.
APPLICATIONS: It has electrical applications due to its high permittivity. It is also used as an opacifier and coloring agent in glazes. Usually it produces a cream color which can be altered to gray. In conjunction with iron the resulting color range is from yellow to brown.

TITANIUM NITRIDE, TiN

MELTING TEMPERATURE: 2930^0 C
SPECIFIC GRAVITY: 5.2
SPECIFIC HEAT: 1.05 kJ/kgK
THERMAL CONDUCTIVITY: 13.4 W/mK at 20^0 C, 26.8 W/mK at 900^0 C
COEFFICIENT OF LINEAR EXPANSION: $9.3X10^{-6}$/K
HARDNESS: 1770 Knoop
RUPTURE STRENGTH: 900 MPa
RUPTURE MODULUS: 235 MPa
COMPRESSIVE STRENGTH: 1,275 MPa
ELECTRICAL RESISTIVITY: $22X10^{-6}$-$30X10^{-6}$ Ohm.cm
COLOR: yellow
SOLUBILITY IN WATER: none
TOXICITY: Titanium nitride is not toxic. However it does decompose and release nitrogen and ammonia which are irritants.
HAZARD: Decomposition of titanium nitride forms ammonia which is combustible.
EMITTANCE: 0.8
STRUCTURE: cubic
APPLICATIONS: It is used as an additive to refractory cement.

TUNGSTEN CARBIDE, WC

MAXIMUM CONTINUOUS SERVICE TEMPERATURE: 550^0 C in oxidizing atmosphere; 2000^0 C in reducing or inert atmosphere
MELTING TEMPERATURE: 2870^0 C

BOILING TEMPERATURE: 6000° C
SPECIFIC GRAVITY: 15.6

SPECIFIC HEAT: 0.18 kJ/kgK at 20° C, 0.21 kJ/kgK at 1100° C
THERMAL CONDUCTIVITY: 29.3 W/mK

COEFFICIENT OF LINEAR EXPANSION: $5.2X10^{-6}$-$7.3X10^{-6}$/K for 20° to 2200° C

HARDNESS: 1900 Knoop for 0° to 600° C, 1500 Knoop at 900° C
MODULUS OF ELASTICITY: 700,000 MPa
TENSILE STRENGTH: 350 MPa

FLEXURAL STRENGTH: 550 MPa at 20° C, 140 MPa at 1650° C
POISSON'S RATIO: 0.26

COMPRESSIVE STRENGTH: 2,750 MPa at 20° C, 1,400 MPa at 1000° C
ABRASION RESISTANCE: excellent

COEFFICIENT OF FRICTION: 0.17 at 20° C, 0.35 at 820° C, 0.40 at 970° C, 0.45 at 1010° C, 0.5 at 1160° C, 0.7 at 1220° C, 1.2 at 1440° C, 1.8 at 1600° C for tungsten carbide on tungsten carbide; 0.15 at 20° C, 0.32 at 800° to 910° C, 0.25 at 1000° C, 0.29 at 1120° C, 0.26 at 1220° C, 0.25 at 1300° to 2030° C for tungsten carbide on graphite; 0.4-0.6 for tungsten carbide on steel; 0.35 for tungsten carbide on copper at 20° C

WELDING ON TEMPERATURE: 1300° C to steel
WEAR RATE: 0.017-0.084 mm.mm/kg.km on steel; in general very wear resistant

ELECTRICAL RESISTIVITY: 10^{-4} Ohm.cm
COLOR: black
SOLUBILITY IN WATER: none
CHEMICAL STABILITY; RESISTANCE TO:
   ACIDS: very good
   BASES: very good

   GASES: It reacts readily with air and slightly with hydrogen above 700° C
TOXICITY: no known health hazard
STRUCTURE: hexagonal
APPLICATIONS: Its high wear resistance finds application in wire-drawing dies. It is also used for high speed drill bits.

TUNGSTEN DISULFIDE, TUNGSTENITE, WS_2

DECOMPOSITION TEMPERATURE: 1250° C
SPECIFIC GRAVITY: 7.5

SPECIFIC HEAT: 0.34 kJ/kgK at 1000° C

THERMAL CONDUCTIVITY: 2.6 W/mK at 750° C, 3.0 W/mK at 1500° C

COEFFICIENT OF LINEAR EXPANSION: $8.5X10^{-6}$/K for 0° to 1500° C
HARDNESS: 1100 Knoop

MAGNETIC SUSCEPTIBILITY: 7.35×10^{-5} m^3/kg
COLOR: brown
SOLUBILITY IN WATER: none
TOXICITY: The ceramic is not known to be toxic but free sulfur combines readily to form toxic compounds.
STRUCTURE: hexagonal

URANIUM CARBIDE, UC, UC$_2$, U$_2$C$_3$

MELTING TEMPERATURE: 2550^0 C for UC, 2350^0-2400^0 C for UC$_2$

BOILING TEMPERATURE: 4370^0 C for UC$_2$

DECOMPOSITION TEMPERATURE: 1730^0 C for U$_2$C$_3$

SPECIFIC GRAVITY: 11.3 for UC$_2$, 13.6 for UC

SPECIFIC HEAT: 0.19 kJ/kgK at 20^0 C, 0.28 kJ/kgK at 1100^0 C

THERMAL CONDUCTIVITY: 23 W/mK over 200^0 to 1000^0 C for UC

COEFFICIENT OF LINEAR EXPANSION: 1.1×10^{-5}/K over 20^0 to 1600^0 C
HARDNESS: 800 Knoop

MODULUS OF ELASTICITY: 200,000 MPa at 20^0 C, 186,000 MPa at 1100^0 C
FLEXURAL STRENGTH: 83 MPa
COMPRESSIVE STRENGTH: 345 MPa

CREEP RATE: 0.002/hr at 1400^0 C and a load of 40 MPa, 0.02/hr at 1500^0 C and a load of 40 MPa
COLOR: gray
SOLUBILITY IN WATER: slight for UC$_2$, UC tends to decompose in water

CHEMICAL STABILITY; RESISTANCE TO:
  ACIDS: tends to decompose
  BASES: tends to decompose
  GASES: reacts with water vapor and carbon dioxide
TOXICITY: radioactive and an established carcinogen
HAZARD: In powdered form it will burn at room temperature. The solid form burns in air above 400^0 C
STRUCTURE: UC is face centered cubic and a stable form up to melting. U$_2$C$_3$ is body centered cubic and remains so until it decomposes. UC$_2$ is a stable tetragonal crystal from about 1500^0 to 1765^0 C. At this temperature it converts to a cubic structure which can form a solid solution with UC.
APPLICATIONS: It is used as a fuel in nuclear reactors.

URANIUM NITRIDE, UN

MELTING TEMPERATURE: 2630^0 C

SPECIFIC GRAVITY: 14.3

SPECIFIC HEAT: 0.18 kJ/kgK at 20° C, 0.22 kJ/kgK at 1100° C

THERMAL CONDUCTIVITY: 23 W/mK for 200° to 1000° C

HARDNESS: 500 Knoop at 20° C, 120 Knoop at 1100° C
MODULUS OF ELASTICITY: 242,000 MPa

FLEXURAL STRENGTH: 76 MPa at 20° C, 131 MPa at 1100° C
SHEAR MODULUS: 97,000 MPa
POISSON'S RATIO: 0.26
COLOR: brown
FLAMMABILITY: It will ignite in the powdered form at ambient temperature.
CHEMICAL STABILITY; RESISTANCE TO:
    ACIDS: poor resistance
    BASES: poor resistance
    GASES: It reacts readily with most gases especially at elevated temperatures.

    METALS: It reacts readily with aluminum above 500° C; with nickel, thorium, titanium and vanadium above 1000° C; with steel above 1300° C; and with liquid sodium and lithium above 950° C.
TOXICITY: radioactive and an established carcinogen
STRUCTURE: cubic
APPLICATIONS: It is used as a fuel in nuclear reactors.

URANIUM OXIDE, UO_2

MAXIMUM CONTINUOUS SERVICE TEMPERATURE: 500° C in oxygen, 1500° C in hydrogen
MELTING TEMPERATURE: 2878° C
SPECIFIC GRAVITY: 11.0

SPECIFIC HEAT: 0.23 kJ/kgK at 25° C, 0.33 kJ/kgK at 1000° C

THERMAL CONDUCTIVITY: 8.5 W/mK at 20° C, 3.6 W/mK for 200° to 1000° C

COEFFICIENT OF LINEAR EXPANSION: 10^{-5}/K for 25° to 1000° C

HARDNESS: 600 Knoop at 20° C, 200 Knoop at 250° C, 130 Knoop at 1000° C

MODULUS OF ELASTICITY: 214,000 MPa at 20° C, 173,000 MPa at 1100° C

TENSILE STRENGTH: 160 MPa for 20° to 500° C, 220 MPa at 700° C, 160 MPa at 1200° to 1400° C

FLEXURAL STRENGTH: 104 MPa at 20° C, 138 MPa at 1100° C

SHEAR MODULUS: 83,000 MPa at 20° C, 55,000 MPa at 1100° C

COMPRESSIVE STRENGTH: 250 MPa at 600° C, 100 MPa at 1400° C

CREEP RATE: 10^{-4}/hr for a load of 48 MPa at 1430° C; 10^{-4}/hr for a load of 14 MPa at 1530° C; 10^{-3}/hr for a load of 41 MPa at 1530° C; 10^{-3}/hr for a load of 7

MPa at 1800° C; 10^{-2}/hr for a load of 35 MPa at 1800° C

ELECTRICAL RESISTIVITY: 10^{13} Ohm.cm at 20° C, 10^{11} Ohm.cm at 500° C

MAGNETIC SUSCEPTIBILITY: 2.96×10^{-5} m^3/kg

ACOUSTIC LONGITUDINAL WAVE VELOCITY: 5.18 km/s

ACOUSTIC LONGITUDINAL WAVE IMPEDANCE: 56.7 kg/m^{2}s

SOLUBILITY IN WATER: none

CHEMICAL STABILITY; RESISTANCE TO:
   ACIDS: reacts
   BASES: reacts

TOXICITY: radioactive and an established carcinogen

STRUCTURE: face centered cubic

APPLICATIONS: It is used as a fuel in nuclear reactors. It is also used as a glaze colorant. Alone the resulting color varies from yellow through orange to black. in conjunction with lead it produces red.

URANIUM SULFIDE, US

MELTING TEMPERATURE: above 2000° C

SPECIFIC GRAVITY: 10.9

SPECIFIC HEAT: 0.28 kJ/kgK

THERMAL CONDUCTIVITY: 11.7 W/mK

COEFFICIENT OF LINEAR EXPANSION: 11.6×10^{-6}/K

MAGNETIC SUSCEPTIBILITY: 3.94×10^{-5} m^3/kg

COLOR: black

CHEMICAL STABILITY; RESISTANCE TO:
   METALS: no reaction except with zirconium

TOXICITY: It is radioactive and an established carcinogen.

APPLICATIONS: It is used as a nuclear fuel.

VANADIUM CARBIDE, VC

MAXIMUM CONTINUOUS SERVICE TEMPERATURE: 700° C in air

MELTING TEMPERATURE: 2684° C

BOILING TEMPERATURE: 3900° C

SPECIFIC GRAVITY: 5.8

SPECIFIC HEAT: 0.51 kJ/kgK at 20° C, 0.88 kJ/kgK at 1100° C

COEFFICIENT OF LINEAR EXPANSION: 6.7×10^{-6}/K for 20° to 1800° C

HARDNESS: 2200 Knoop at 0° C, 1700 Knoop at 200° C, 1100 Knoop at 600° C, 400 Knoop at 900° C

MODULUS OF ELASTICITY: 414,000 MPa at 20° C, 331,000 MPa at 1900° C, 255,000 MPa at 2200° C

SHEAR MODULUS: 159,000 MPa
POISSON'S RATIO: 0.22
COMPRESSIVE STRENGTH: 607 MPa
ELECTRICAL RESISTIVITY: 10^{-4} Ohm.cm
COLOR: black
SOLUBILITY IN WATER: none
CHEMICAL STABILITY; RESISTANCE TO:
   GASES: it is attacked by hydrogen bearing gases above 1350° C
TOXICITY: It is an irritant to the respiratory system when it is inhaled in dust form.
STRUCTURE: cubic
APPLICATIONS: It is used in carbide cutting tools.

VANADIUM TRIOXIDE, V_2O_3

MELTING TEMPERATURE: 1970° C
SPECIFIC GRAVITY: 4.87
SPECIFIC HEAT: 0.55 kJ/kgK at 20° C, 0.86 kJ/kgK at 1100° C
MAGNETIC SUSCEPTIBILITY: $2.48X10^{-5}$ m^3/kg
COLOR: black
SOLUBILITY IN WATER: slight
CHEMICAL STABILITY; RESISTANCE TO:
   METALS: attacked by liquid lithium
TOXICITY: It is an irritant to the respiratory system when inhaled in dust form.
NEEL TEMPERATURE: 170 K
STRUCTURE: hexagonal
APPLICATIONS: It is used as a flux in glazes and as a yellow pigment in combination with tin or zirconia.

XENON, Xe

MELTING TEMPERATURE: 161 K
BOILING TEMPERATURE: 165 K
SPECIFIC GRAVITY: 0.0059
SPECIFIC HEAT: 0.158 kJ/kgK
THERMAL CONDUCTIVITY: 0.05 W/mK
MAGNETIC SUSCEPTIBILITY: $-4.20X10^{-9}$ m/kg
COLOR: colorless gas
CHEMICAL STABILITY: Xenon is inert except for reactions with fluorine and polymers.
TOXICITY: It does interfere with the assimilation of oxygen in the lungs. In addition, xenon compounds are very chemically active and react with the body.
STRUCTURE: face centered cubic

APPLICATIONS: It is used for speciality lighting.

YTTERBIUM OXIDE, YTTERBIA, Yb_2O_3

MELTING TEMPERATURE: 2333° C
SPECIFIC GRAVITY: 9.17
SPECIFIC HEAT: 0.3 kJ/kgK at 20° C, 0.33 kJ/kgK at 1650° C
HARDNESS: 650 Knoop
MODULUS OF ELASTICITY: 186 MPa

FLEXURAL STRENGTH: 76 MPa for 20° to 1200° C
SHEAR MODULUS: 76,000 MPa
POISSON'S RATIO: 0.28
COMPRESSIVE STRENGTH: 228 MPa
COLOR: clear
SOLUBILITY IN WATER: none
TOXICITY: Its radioactivity causes a general health hazard but it may also have beneficial medical applications.
STRUCTURE: cubic
APPLICATIONS: It is used as an additive in electrically conductive ceramics.

YTTRIUM OXIDE, Y_2O_3

MELTING TEMPERATURE: 2410°
SPECIFIC GRAVITY: 5.0
SPECIFIC HEAT: 0.46 kJ/kgK at 20° C, 0.55 kJ/kgK at 550° C, 0.59 kJ/kgK at 1400° C

THERMAL CONDUCTIVITY: 14 W/mK at 20° C, 4W/mK at 550° C, 3 W/mK at 1650° C, 6 W/,nK at 2200° C

COEFFICIENT OF LINEAR EXPANSION: $8.1X10^{-6}$/K for 20° to 650° C, $9.4X10^{-6}$/K for 650° to 1500° C

HARDNESS: 770 Knoop at 20° C, 170 Knoop at 1100° C

MODULUS OF ELASTICITY: 159,000 MPa at 20° C, 131,000 MPa at 1350° C

FLEXURAL STRENGTH: 104 MPa at 20° C, 138 MPa at 1000° C, 124 MPa at 1500° to 1750° C

POISSON'S RATIO: 0.3 for 20° to 1000° C
COMPRESSIVE STRENGTH: 186 MPa
MAGNETIC SUSCEPTIBILITY: $5.58X10^{-7}$ m^3/kg
REFRACTIVE INDEX: 1.92
TRANSMISSION OF LIGHT: 10% or more through a 2 mm thickness in the region 0.26-9.2 microns

COLOR: clear to yellow
SOLUBILITY IN WATER: 0.0002 g/100ml
CHEMICAL STABILITY; RESISTANCE TO:

METALS: good resistance to lithium up to 400° C
TOXICITY: Its radioactivity causes a general health hazard but it may also have beneficial medical applications which have not been fully explored.
STRUCTURE: cubic
APPLICATIONS: It is used as a stabilizer in zirconia, a microwave filter and a component in television tubes.

ZINC OXIDE, ZINCITE, ZnO

MELTING TEMPERATURE: 1975° C
SPECIFIC GRAVITY: 5.6
SPECIFIC HEAT: 0.5 kJ/kgK
HARDNESS: 4 Mohs
MAGNETIC SUSCEPTIBILITY: -5.78×10^{-7} m^3/kg
REFRACTIVE INDEX: 2.0
COLOR: white
ACOUSTIC LONGITUDINAL WAVE VELOCITY: 6.40 km/s
ACOUSTIC SHEAR WAVE VELOCITY: 2.95 km/s
ACOUSTIC LONGITUDINAL WAVE IMPEDANCE: 36.4 kg/m^2s
SOLUBILITY IN WATER: 0.0002 g/100ml
TOXICITY: A fungicide, it is also used as a dietary supplement which does not add confidence to prepared mineral pills.
STRUCTURE: hexagonal
APPLICATIONS: It is used as an additive to glazes because it reduces the tendency for crazing and increases the resistance to scratching. It also serves as an opacifier.

ZINC SELENIDE, ZnSe

MELTING TEMPERATURE: 1520° C
SPECIFIC GRAVITY: 5.42
SPECIFIC HEAT: 0.34 kJ/kgK at 20° C, 0.4 kJ/kgK at 800° C
THERMAL CONDUCTIVITY: 11 W/mK
COEFFICIENT OF LINEAR EXPANSION: 7.5×10^{-6}/K
THERMAL SHOCK: poor resistance
HARDNESS: 105 kg/mm^2
FLEXURAL STRENGTH: 60 MPa
MODULUS OF ELASTICITY: 69,000 MPa
RUPTURE MODULUS: 48 MPa
DIELECTRIC CONSTANT: 9 at 20° C, 9.9 at 500° C

REFRACTIVE INDEX: 2.89
COLOR: yellow to red
SOLUBILITY IN WATER: none
TOXICITY: This is a moderate fire and explosion hazard as well as a recognized carcinogen due to the selenium.
STRUCTURE: cubic
APPLICATIONS: It is used as an infrared optical window.

ZINC SULFIDE, WURTZITE, SPHALERITE, ZnS

MELTING TEMPERATURE: 1700° C
SPECIFIC GRAVITY: 3.9 for hexagonal, 4.1 for cubic
SPECIFIC HEAT: 0.5 kJ/kgK
THERMAL CONDUCTIVITY: 11 W/mK
COEFFICIENT OF LINEAR EXPANSION: 6.7×10^{-6}/K
THERMAL SHOCK: fair resistance
HARDNESS: 3.5-4 Mohs
TENSILE STRENGTH: 60 MPa from 20° to 600° C
MODULUS OF ELASTICITY: 97,000 MPa
RUPTURE MODULUS: 97 MPa
POISSON'S RATIO: 0.3
REFRACTIVE INDEX: 2.25
COLOR: clear
OPTICAL TRANSMISSION: 75% for 3 to 9 microns over 20° to 500° C for a 5 mm thickness
DIELECTRIC CONSTANT: 8.8 at 20° C, 9.9 at 600° C
SOLUBILITY IN WATER: slight
TOXICITY: The sulfur component is a skin irritant. Zinc is benign.
STRUCTURE: Sphalerite or beta form is cubic and transforms to wurtzite or alpha form which is hexagonal. The transformation takes place at 1020° C.
APPLICATIONS: It is used in the making of x-ray and television tubes. It is also used as a fungicide.

ZIRCONIUM CARBIDE, ZrC

MELTING TEMPERATURE: 3400°-3540° C
BOILING TEMPERATURE: 5100° C
SPECIFIC GRAVITY: 6.56-6.73
SPECIFIC HEAT: 0.2 kJ/kgK at 20° C, 0.56 kJ/kgK at 1100° C
THERMAL CONDUCTIVITY: 2.0 W/mK at 20° C, 43.6 W/mK at 2100° C, 46.9 W/mK at 2200° C
COEFFICIENT OF LINEAR EXPANSION: 6.7×10^{-6}/K for 20° to 600° C

THERMAL SHOCK: Resistance is never good for a carbide but zirconium carbide is superior to both tantalum carbide and titanium carbide.

HARDNESS: 2000 Knoop at 0° C, 2100 Knoop at 20° C, 1000 Knoop at 400° C, 300 Knoop at 900° C

MODULUS OF ELASTICITY: 380,000 MPa at 20° C, 310,000 MPa at 2200° C

TENSILE STRENGTH: 106 MPa at 20° to 1200° C, 69 MPa at 1600° C, 35 MPa at 2000° C

FLEXURAL STRENGTH: 173 MPa at 20° C, 207 MPa at 1200° C

SHEAR MODULUS: 173,000 MPa

POISSON'S RATIO: 0.2

COMPRESSIVE STRENGTH: 2,000 MPa at 20° C, 483 MPa at 1000° C, 276 MPa at 1200° C

CREEP RATE: 0.05/hr at 1800° C and a load of 14 MPa

WEAR RATE: 0.1 mm.mm/kg.km on steel

ELECTRICAL RESISTIVITY: $4.2X10^{-5}$-$6.7X10^{-5}$ Ohm.cm at 20° C, $1.4X10^{-4}$ Ohm.cm at 980° C

MAGNETIC SUSCEPTIBILITY: $-3.27X10^{-7}$ m^3/kg

COLOR: gray

SOLUBILITY IN WATER: none

CHEMICAL STABILITY; RESISTANCE TO:
ACIDS: poor to fair resistance
BASES: poor to fair resistance

GASES: It reacts readily with air above 1150° C. It also has a slight reactivity with carbon monoxide and nitrogen above 1500° C

TOXICITY: In the dust form it is a lung irritant that should be avoided.

STRUCTURE: face centered cubic with a lattice parameter of 4.69 Angstroms

APPLICATIONS: It is used in cutting tools and as an abrasive.

ZIRCONIUM OXIDE, ZIRCONIA, BADDELEYITE, ZIRKITE, FAVAS, CUBIC ZIRCONIA OR PARTIALLY STABILIZED ZIRCONIA (PSZ), FULLY STABILIZED ZIRCONIA (FSZ), ZrO_2

MAXIMUM CONTINUOUS SERVICE TEMPERATURE: 2200° C for oxidizing atmosphere

MELTING TEMPERATURE: 2700° C for baddeleyite, 2950° C for zirconia

BOILING POINT: about 5000° C

MINIMUM MELTING POINT OF BINARY SYSTEMS: 1710° C for alumina; 2145° C for beryllia; 2140° C for calcium oxide; 2113° C for magnesium oxide; 1675° C for silica; 2680° C for thoria; 1750° C for titania; 2200° C for strontium oxide

SPECIFIC GRAVITY: 5.89 for baddeleyite; 5.72 for alpha zirconia; 5.31 for beta zirconia; 5.3 for partially stabilized zirconia

SPECIFIC HEAT: 0.45 kJ/kgK at 20° C, 0.67 kJ/kgK at 1100° C, 0.70 kJ/kgK at 2200° C

THERMAL CONDUCTIVITY: 2 W/mK for 0° to 1200° C

COEFFICIENT OF LINEAR EXPANSION: 9.5×10^{-6}/K for zirconia; 7.6×10^{-6}/K at 20° to 420° C, 12.0×10^{-6}/K at 420° to 750° C, 15.5×10^{-6}/K at 750° to 2200° C for FSZ

THERMAL SHOCK: moderately good

HARDNESS: 1160 Knoop for zirconia; 1520 Knoop for FSZ; 6.5 Mohs for baddeleyite

MODULUS OF ELASTICITY: 242,000 MPa at 20° C, 214,000 MPa at 550° C, 152,000 MPa at 1200° C for zirconia; 200,000 MPa for partially stabilized zirconia; 428,000 MPa for zirconia whisker

TENSILE STRENGTH: 4,140 MPa for zirconia whisker; 69 MPa at 20° C, 83 MPa at 1200° C, 7 MPa at 1530° C for PSZ

FLEXURAL STRENGTH: 1,000 MPa at 20° C, 138 MPa at 500° C, 117 MPa at 1000° C, 69 MPa at 1200° C for partially stabilized zirconia. These values are increased by about 20% for the fully stabilized form.

FRACTURE TOUGHNESS: 8 MPa.m$^{1/2}$ for partially stabilized zirconia

SHEAR STRENGTH: 186 MPa for zirconia

SHEAR MODULUS: 104,000 MPa at 20° C, 66,000 MPa at 1100° C for zirconia

POISSON'S RATIO: 0.27 for zirconia; 0.23-0.32 for PSZ

COMPRESSIVE STRENGTH: 2,000 MPa at 20° C, 1,600 MPa at 500° C, 1,170 MPa at 1000° C, 124 MPa at 1400° C for PSZ

CREEP RATE: 3×10^{-5} /hr at 1300° C and a load of 12 MPa

WEAR RATE: 0.01-0.085 mm.mm/kg.km on steel

WEIBULL MODULUS: 14

ELECTRICAL RESISTIVITY: 10^{14} Ohm.cm at 200° C, 10^6 Ohm.cm at 700° C

DIELECTRIC STRENGTH: 5 kv/mm

DIELECTRIC CONSTANT: 12 at 1 MHz

DISSIPATION FACTOR: 0.12 at 1 MHz

MAGNETIC SUSCEPTIBILITY: -1.73×10^{-7} m^3/kg

REFRACTIVE INDEX: 2.1-2.2 for baddeleyite

COLOR: clear to yellow for baddeleyite, white for zirconia, clear for PSZ

SOLUBILITY IN WATER: none

CHEMICAL STABILITY; RESISTANCE TO:

ACIDS: not very resistant

BASES: good resistance

GASES: readily reacts with hydrogen, water vapor and ammonia; limited reaction with sulfur and carbon containing gases

METALS: good resistance to metals in solid and liquid state except lithium

TOXICITY: In the dust form it is a lung irritant that should be avoided.

EMITTANCE: 0.6

STRUCTURE: Baddeleyite (or beta zirconia when in the pure state rather than as a mineral) is monoclinic and the stable form at low temperature. It transforms to the alpha phase which is tetragonal at 1000° C with a correspondingly large increase in volume. On cooling the reverse transformation is usually delayed until 625° C. There is also a form which is simple cubic. This is not stable unless some impurities are present. The cubic phase has the advantage of being stable over the entire temperature range so it does not experience a large volume change on transformation.

ADDITIVES: About 5% magnesia and calcium oxide are added to zirconia and it is heated to 1700° C to render the stable form of cubic zirconia. Calcium oxide enhances the resistance of the material to bases in all forms of zirconia. For fiber formation, yttria is the preferred stabilizer to produce a cubic zirconia.

APPLICATIONS: The cubic form is the only phase used widely for structural use since its volume is stable. It is used as a containment vessel for melting metals in a vacuum, wire-drawing dies and injection forming equipment. The fiber form is used for heat resistant clothing and as a reinforcement to composites of all types. zirconium oxide is also used to aid in color development and as an opacifer in glazes.

ZIRCONIUM SILICATE, ZIRCON, $ZrO_2 \cdot SiO_2$

DECOMPOSITION TEMPERATURE: 1500°-1800° C it transforms to zirconia plus silica. If the transformation can be suppressed then melting takes place at 2550° C.

SPECIFIC GRAVITY: 4.2-4.6

SPECIFIC HEAT: 0.54 kJ/kgK

THERMAL CONDUCTIVITY: 6 W/mK

COEFFICIENT OF LINEAR EXPANSION: $4X10^{-6}$/K

THERMAL SHOCK RESISTANCE: good

HARDNESS: 7-8 Mohs

TENSILE STRENGTH: 96 MPa

COMPRESSIVE STRENGTH: 524 MPa

ELECTRICAL RESISTIVITY: 10^{14} Ohm.cm

DIELECTRIC STRENGTH: 10 kv/mm

DIELECTRIC CONSTANT: 9 at 1 MHz

DISSIPATION FACTOR: 0.01 at 1 MHz

REFRACTIVE INDEX: 1.95

BIREFRINGENCE: 0.055

SOLUBILITY IN WATER: none

TOXICITY: The toxicity is primarily due to the silica content.

STRUCTURE: tetragonal

APPLICATIONS: It finds use as a refractory. In the glass industry it is used as a color stabilizer and an opacifier in glazes.

GLOSSARY

ABBE NUMBER - is the reciprocal value of the relative dispersion. It is calculated by the formula $(n_D - 1)/(n_F - n_C)$ This is the characteristic dispersion at the wavelengths of C=656 nm, D=589 nm and F=486 nm.

ACANTHITE - is the mineral form of Ag_2S. It has a rhombic structure, a specific gravity of 7.3, a transformation temperature of 175° C, and a very slight solubility in water.

ACTINOLITE - is a type of asbestos. See the data section under asbestos for the properties.

ADAMITE - is the mineral form of $Zn_3(AsO_4)_2·Zn(OH)_2$. It has a rhombic structure, a specific gravity of 4.5 and a decomposition temperature of 250° C.

AGGREGATE - is any particulate matter ranging from fine sand to course stones of 40 mm. diameter included in concrete. The preferred shape is spherical because it packs better. Also, a distribution of size is desirable so that the smaller spheres may fill in around the larger ones.

ALABANDITE - is the mineral form of MnS. It has a cubic structure, a specific gravity of 4.0 and a solubility in water of 0.005 g/100ml.

ALAMOSITE - is the mineral form of $PbSiO_3$. It has a monoclinic structure, a specific gravity of 6.5, a melting point of 766° C. and no solubility in water.

ALBITE - is the mineral form of $Na_2O·Al_2O_3·6SiO_2$. It has a triclinic structure, a specific gravity of 2.6, an index of refraction of 1.5, a birefringence of 0.008, a melting point of 1100° C. and a hardness of 6.0-6.5 Mohs. It is a component of feldspar.

ALITE - is the mineral form of $3CaO·SiO_2$. It has a monoclinic structure, an index of refraction of 1.7 and a melting point of approximately 1900° C.

ALKALI FELDSPAR - is composed mainly of potassium feldspar and albite.

ALLOPHANE - is a general term used for non-crystalline solutions of silica, alumina and water. Usually it is $Al_2O_3SiO_2·5H_2O$

ALTAITE - is the mineral form of PbTe. It has a cubic structure, a specific gravity of 8.2 and a melting point of 917° C.

ALUMINA (alpha form) - is the commercially important form of aluminum oxide. See the data section for the properties.

ALUMINA (beta form) - is $Na_2O \cdot 11Al_2O_3$. It was miss-named because it was originally mistaken for pure aluminum oxide.

ALUMINA (gamma form) - is the hexagonal closed packed form of aluminum oxide. See the data section for the properties.

ALUMINO-SILICATE - are alumina and silica containing ceramics.

ALUMINUM BROMIDE - $AlBr_3$. It has a rhombic structure, a specific gravity of 2.64, a melting point of 97.5^0 C. and a boiling point of 263.3^0 C.

ALUMINUM BORIDE - AlB_{12}. See the data section for the properties.

ALUMINUM FLUOSILICATE - $2AlFO \cdot SiO_2$. The mineral form is topaz.

ALUMINUM NITRIDE - AlN. See the data section for properties.

ALUMINUM OXIDE - Al_2O_3. See the data section for properties.

ALUMINUM OXIDE MONOHYDRATE - $Al_2O_3 \cdot H_2O$. The mineral forms are bauxite, boehmite and diaspore.

ALUMINUM OXIDE TRIHYDRATE - $Al_2O_3 \cdot 3H_2O$. Gibbsite and bayerite are the mineral forms which change phase to boehmite on heating.

ALUMINUM SILICATE - $Al_2O_3 \cdot SiO_2$. See the data section for the properties.

ALUMINUM SULFATE HYDRATE - $Al_2(SO_4)_3 \cdot 18H_2O$. The mineral form is alunogenite.

ALUNOGENITE - is the mineral form of aluminum sulfate hydrate. It has a monoclinic structure, a specific gravity of 1.69, a melting point of 86.5^0 C and an index of refraction of 1.47.

AMBLYGONITE - $2LiF \cdot Al_2O_3 \cdot P_2O_5$. It has a triclinic structure, a specific gravity of 3.0-3.1 and a hardness of 6 Mohs. It is used as a flux and regulator in glazes as well as an agent to reduce vitrification temperature and porosity in porcelains.

216

AMMONIUM DICHROMATE - $(NH_4)_2Cr_2O_7$. It has a specific gravity of 2.2, a melting point of 170° C., a solubility in cold water of 31 g/100ml. and in hot water of 89 g/100ml. It is used as a coloring agent in ceramics. If in the presence of tin, calcium or silica it produces pink. If it is in the presence of zinc, brown results whereas in the presence of lead yellow results. If none of the above are present than the color will be green.

AMOSITE - is a type of asbestos. See the data section under asbestos for the properties.

ANATASE - is a mineral form of titanium dioxide. See the data section for properties.

ANAUXITE - is a clay mineral with the formula $Al_2O_3 \cdot 3SiO_2 \cdot 2H_2O$ and a thin hexagonal plate structure.

ANDALUSITE - is a mineral form of aluminum silicate. See the data section for the properties.

ANGLESITE - is the mineral form of $PbSO_4$. It has a monoclinic or rhombic structure, an index of refraction of 1.9, a specific gravity of 6.2, a melting point of 1170° C, a solubility in cold water of 0.0043 g/100ml and in hot water of 0.0056 g/100ml.

ANHYDRITE - is the mineral form of calcium sulfate. It has a triclinic, monoclinic or rhombic structure, a specific gravity of 2.61-2.96, a hardness of 6.0-6.5 Mohs, an index of refraction of 1.5-1.6, a melting point of 1450° C., a solubility in cold water of 0.209 g/100ml and in hot water of 0.1619 g/100ml.

ANORTHITE - $CaO \cdot Al_2O_3 2SiO_2$. It has a triclinic structure, an index of refraction of 1.58, a birefringence of 0.008, a specific gravity of 2.76, a hardness of 6-6.5 Mohs, a coefficient of thermal expansion of $4.5X10^{-6}$/C over 100-200 C and a melting point of 1551° C. It is the limespar component of feldspar.

ANTHOPHYLLITE - is a form of asbestos. See the data section under asbestos for the properties.

ANTIMONY OXIDE - is an alternate term for antimony trioxide.

ANTIMONY TETRAOXIDE - Sb_2O_4. The mineral form is cerantite.

ANTIMONY TRIOXIDE - Sb_2O_3. See the data section for the properties.

ANTIMONY TRISULFIDE - Sb_2S_3. The mineral form is stibnite.

APATITE - is a mineral of the composition $Ca_4(CaF)(PO_4)_3$ or $Ca_4(CaCl)(PO_4)_3$. It has a specific gravity of 3.2 for the crystalline form and 3.0 for the glassy form, and the melting point varies from 1300° C to 1650° C depending on the composition.

ARAGONITE - is a mineral form of calcium carbonate. See the data section for the properties.

ARCANITE - is the mineral form of K_2SO_4. It has a rhombic or hexagonal structure, an index of refraction of 1.5, a specific gravity of 2.7, a solubility in cold water of 12 g/100ml. and in hot water of 24 g/100ml. A phase change occurs at 588° C., melting occurs at 1069° C. and boiling occurs at 1689° C.

ARGENTITE - is the mineral form of Ag_2S. It has a cubic structure, a specific gravity of 7.3, a melting point of 825° C and a solubility in water of 0.8 g/100 ml.

ARSENIC - is used as a fining agent in glass making. During processing it gives off oxygen which bubbles though the glass and sweeps out seeds.

ARSENIC DISULFIDE - As_2S. The mineral form is realgar.

ARSENIC TRIOXIDE - As_2O_3. See the data section for the properties.

ARSENIC TRISULFIDE - As_2S_3. The mineral form is orpiment. See the data section for the properties.

ARSONOLITE - is the mineral form of arsenic trioxide. See the data section for the properties.

ARTINITE - is the mineral form of $MgCO_3Mg(OH_2) \cdot 3H_2O$. It has a rhombic structure, an index of refraction of 1.5 and a specific gravity of 2.0.

ASBESTOS - is a general term for a silicate which forms fibers. The properties are given in the data section.

ASCHARITE - is the mineral form of $Mg_2B_2O_4 \cdot H_2O$. It has an orthorhombic structure, an index of refraction of 1.5, and a specific gravity of 2.6-2.7.

ATACAMITE - is the mineral form of $Cu_2(OH)_3Cl$. It has an orthorhombic structure and a specific gravity of 3.8.

ATTAPULGITE - is an unusual clay which is fibrous in structure. The theorized formula is $(OH_2)_4(OH)_5Mg_5Si_8O_{20} \cdot 4H_2O$.

AVOGADRITE - is the mineral form of KBF_4. It has a rhombic or cubic structure, an index of refraction of 1.3, a specific gravity of 2.5, a decomposition temperature of 350° C., a solubility on cold water of 0.4 g/100ml and in hot water of 6.3 g/100ml.

AZURITE - is the mineral form of $2CuCO_3Cu(OH)_2$. It has a monoclinic structure, an index of refraction of 1.7-1.8, a specific gravity of 3.88, a decomposition temperature of 220° C and no solubility in water.

BADDELEYITE - is the mineral form of zirconium oxide. See the data section for properties.

BALL CLAY - is a sedimentary plastic clay which is dark due to organic impurities. Kaolinite is a major constituent. The name comes from the English practice of cutting the clay into balls during mining. The final product will burn to be white. Its primary use is in making stonewear dishes.

BARITE - is the mineral form of barium sulfate. It has a rhombic structure which transforms to monoclinic at 1149° C., an index of refraction of 1.6, a specific gravity of 4.5, a melting point of 1580° C., and a solubility in water of 0.0003 g/100ml. It is used as a source of barium carbonate.

BARIUM ANORTHITE - is an alternate term for celsian.

BARIUM CARBONATE - $BaCO_3$. The mineral form, witherite, converts to beta barium carbonate at 811° C. Beta barium carbonate converts to alpha barium carbonate, which is hexagonal, at 982° C. Alpha barium carbonate breaks down to barium oxide at 1450° C. The solubility in water for all three forms is 0.002 g/100ml at 20° C and 0.006 g/100ml at 100° C. The specific gravity for all three forms is 4.43. It is used as a flux in glazes, in conjunction to deflucculants in clay and as an additive to bricks to prevent efflorescence.

BARIUM NITRATE - $Ba(NO_3)_2$. The mineral form is nitrobarite.

BARIUM OXIDE - BaO. See the date section for properties.

BARIUM SULFATE - $BaSO_3$. The mineral form is barite.

BARIUM TITANATE - $BaTiO_2$. It has a specific gravity of 5.8, and a Young's Modulus of 135,000 MPa. Barium titanate is used as a capacitor transducer.

BARYSILITE - is the mineral form of $Pb_2Si_2O_7$. It has an index of refraction of 2.1, a specific gravity of 6.7, a trigonal structure and no solubility in water.

219

BARYTA SPAR - is an alternate term for celsian.

BARYTE - is an alternate spelling for barite.

BASTNAESITE - is the mineral form of $CeFCO_3$. It has a hexagonal structure, an index of refraction of 1.7-1.8 and a specific gravity of 5.0.

BAUXITE - is the mineral form of aluminum oxide monohydrate. It has a mono-clinic structure, a specific gravity of 2.42 and an index of refraction of 1.6. It is the same chemical composition as boehmite and diaspore.

BAYERITE - is mineral form of aluminum oxide trihydrate with the same chemical composition as gibbsite and hydra-argilite. It has a specific gravity of 2.53 and transforms to boehmite at high temperature.

BEIDELLITE - is a clay mineral of the form $Al_2O_3 \cdot 3SiO_2 x H_2O$.

BELLINGERITE - is the mineral form of $CU(IO_3)_2H_2O$. It has a triclinic structure, a specific gravity of 4.9, loses the water component at 248° C and decomposes at 290° C. It has a solubility in cold water of 0.33 g/100ml and in hot water of 0.65g/100ml.

BENITOITE - $BaTi(SiO_3)_3$

BENTONITE - is a clay derived from volcanic ash; the major component is mont-morillonite. It is used as a plasticiser for other clays, especially ball clays.

BERTRANDITE - is the mineral form of $Be_4Si_2O_7(OH)_2$. It has a rhombic struc-ture, an index of refraction of 1.6 and a specific gravity of 2.6.

BERYL - is the mineral form of $Be_3Al_2(SiO_3)_6$. It has a hexagonal structure, an index of refraction of 1.5, a specific gravity of 2.66, a hardness of 7.5-8 Mohs and a melting point of 1410° C. It is the primary source of beryllium oxide and is used as a flux in some electrical porcelains.

BERYLLIA - is a contracted term for beryllium oxide.

BERYLLIUM ALUMINATE - $BeO \cdot Al_2O_3$. The mineral form is chrysoberyl.

BERYILLIUM OXIDE - BeO. See the data section for properties.

BISCHOFITE - is the mineral form of $MgCl_2 \cdot 6H_2O$. It has a monoclinic structure, an index of refraction of 1.5, a specific gravity of 1.6, a decomposition temperature of 116° C, a solubility in cold water of 167 g/100ml and in hot water of 367 g/100ml.

BISMUTHGLANCE - is an alternate term for bismuthinite.

BISMUTHINITE - is the mineral form of bismuth trisulfide, It has a rhombic structure, an index of refraction of 1.3-1.5, a specific gravity of 7.39, decomposes at 685° C and a solubility in cold water of 0.000018 g/100ml.

BISMUTH OXIDE - Bi_2O_3. It has a specific gravity of 8.2-8.9, a melting point of 820° C, and no solubility in water. Bismuth oxide is used as a flux in glazes.

BISMUTH SILICATE - $2Bi_2O_3 \cdot 3SiO_2$. The mineral form is eulytite.

BISMUTH TRISULFIDE - Bi_2S_3. The mineral form is bismuthinite.

BLOATING - is the blistering produced when clay is overfired.

BLOEDITE - is the mineral form of $Na_2SO_4MgSO_4 \cdot 4H_2O$. It has a monoclinic structure, an index of refraction of 1.5, a specific gravity of 2.2 and a slightly solubility in water.

BLUE VERDIGRIS - $Cu(C_2H_3O_2)_2 \cdot CuO \cdot 6H_2O$. It is a greenish blue powder which is slightly soluble in water and is used as a pigment.

BLUE VITRIOL - is an alternate term for chalcanthite.

BOBIERITE - is the mineral form of $Mg_3(PO_4)_2 \cdot 8H_2O$. It has a monoclinic structure, an index of refraction of 1.5, a specific gravity of 2.2 and a slightly solubility in water.

BODY - is a mixture of clay, water and other minerals made to a workable consistency for manufacture.

BOEHMITE - is the mineral form of aluminum oxide monohydrate. It has an orthorombic structure, a specific gravity of 3.01 and the same chemical composition as bauxite and diaspore.

BONE ASH - is the result of thoroughly calcinated bones. It has an approximate composition of $4Ca_3(PO_4)_2CaCO_3$ and is used in bone china.

BONE CHINA (BRITISH) - is a composite of anorthite and whitlockite in a matrix of phosphate glass.

221

BORAX - is the mineral form of $Na_2O \cdot 2B_2O_3 \cdot 10H_2O$. It has a monoclinic structure, an index of refraction of 1.5, a specific gravity of 1.7, a solubility at 0° C of 2.3 g/100ml, at 20° C of 5.1 g/100ml and at 100° C of 191 g/100ml. It finds use as a flux in glazes.

BORIC OXIDE - is an alternate term for boron oxide.

BOROCALCITE - is an alternate term for pandermite.

BORON CARBIDE - B_4C. See the data section for properties.

BORON NITRIDE - BN. See the data section for properties.

BORON OXIDE - B_2O_3. It is used in glass making to render the glass less viscose during processing. As a flux in glazes it can replace lead. It has a rhombic structure, an index of refraction of 1.61, a specific gravity of 2.46, a melting point of 450° C and a boiling point about 1860° C. Its solubility in water is 1.1 g/100ml at 0° C and 15.7 g/100ml at 10° C. The glass form of boron oxide permits the transmission of X-rays through it.

BRAUNITE - is the mineral form of Mn_2O_3. It has a cubic structure, a specific gravity of 4.5 and no solubility in water.

BREITHAUPTITE - is the mineral form of NiSb. It has an hexagonal structure, a specific gravity of 7.5, a melting point of 1158° C and decomposition point of 1400° C.

BREUNNERITE - $(Mg,Fe)CO_3$.

BRICK CLAY - contains mainly kaolinite and illite clays with some iron oxides to give it its characteristic red color. Besides building bricks, it finds use in stonewear and earthenwear.

BRIDGING - is the term used for an atom that continues the network structure in a glass. This is usually an oxygen atom.

BROCHANTITE - is the mineral form of $CuSO_4 \cdot 3Cu(OH)_2$. It has a monoclinic structure, an index of refraction of 1.7-1.8, a specific gravity of 3.8, a decomposition temperature of 300° C and no solubility in water.

BROMELLITE - is the mineral form of berillium oxide. See the data section for properties.

BROMYRITE - is the mineral form of silver bromide. See the data section for the properties.

BROOKITE - is a mineral form of titanium dioxide. See the data section for the properties.

BRUCITE - is the mineral for of $Mg(OH)_2$. It has a layered hexagonal structure, an index of refraction of 1.6, a specific gravity of 2.4, a solubility in cold water of 0.001 g/100ml and in hot water of 0.004 g/100ml.

BRUNSWICK GREEN - $CuCl_2 \cdot 3CuO \cdot 4H_2O$, is a green pigment.

BRUSHITE - is the mineral form of $CaHPO_4 \cdot 2H_2O$. It has a triclinic structure, an index of refraction of 1.5, a specific gravity of 2.3, a solubility in cold water of 0.03 g/100ml and in hot water of 0.08 g/100ml.

BUNSENITE - is the mineral form of NiO. It has a cubic structure, a specific gravity of 6.7, a melting point of 1984° C and no solubility in water.

CADMIUM CHLORIDE - $CdCl_2$. See the data section for the properties.

CADMIUM SULFIDE - CdS. See the data section for the properties.

CALCAREOUS CEMENT - is the common form of cement used in construction.

CALCIA - is an alternate term for calcium oxide.

CALCITE - is a mineral form of calcium carbonate. See the data section for the properties.

CALCIUM CARBONATE - $CaCO_3$. See the data section for the properties.

CALCIUM FELDSPAR - is an alternate term for anorthite.

CALCIUM FLUORIDE - CaF_2. See the data section for the properties.

CALCIUM IODATE - $Ca(IO_3)_2$. The mineral form is lautarite.

CALCIUM MAGNESIUM CARBONATE - $CaCO_3 \cdot MgCO_3$. The mineral form is dolomite.

CALCIUM MAGNESIUM METASILICATE - $CaO \cdot MgO \cdot 2SiO_2$. The mineral form is diopside.

CALCIUM MAGNESIUM ORTHOSILICATE - $3CaO \cdot MgO \cdot SiO_2$. The mineral form is mervinite.

CALCIUM MOLYBDATE - $CaMoO_4$. The mineral form is powellite.

CALCIUM OXIDE - CaO. It has a cubic structure, an index of refraction of 1.84, a specific gravity of 3.3-3.4, a melting point of 2614° C, a boiling point of 2850° C., a solubility in cold water of 0.131 g/100ml and in hot water of 0.07 g/100ml.

CALCIUM SULFATE - $CaSO_4$. The mineral form is anhydrite.

CALCIUM SULFIDE - CaS. The mineral form is oldhamite.

CAPILLARIES - refers to the space between particles in a ceramic. Usually this term in applied to clay and water in its greenware state.

CARBONIED FIBER - is a fiber with a carbon content between 91% and 98%. Below this amount it is partially carbonized fiber and above this amount it is graphite fiber.

CARNALITE - is the mineral form of $KCl \cdot MgCl_2 \cdot 6H_2O$. It has a rhombic structure, an index of refraction of 1.5, a specific gravity of 1.6, a melting point of 265° C and a solubility in water of 65 g/100ml.

CARNEGIEITE - is the mineral form of $NaAlSiO_4$. It can be formed from heating nepheline to 1254° C.

CASSEL YELLOW - $PbCl_2 \cdot 7PbO$. It is a yellow pigment which is insoluble in water.

CASSITERITE - is the mineral form of tin dioxide. See the data section for the properties.

CARAPLEITE - $Na_2ZrSi_3O_9 \cdot 2H_2O$

CELESTITE - is the mineral form of strontium sulfate. It has a rhombic structure, an index of refraction of 1.6, a specific gravity of 4.0, a melting point of 1605° C, a solubility in cold water of 0.0113 g/100ml and in hot water of 0.014 g/100ml. It is used as a source for strontium carbonate.

CELSIAN - $BaO \cdot Al_2O_3 \cdot 2SiO_2$. It has a triclinic structure, a specific gravity of 3.37, a coefficient of thermal expansion of $2.7 \times 10^{-6}/C$ over 20° to 100° C, and a hardness of 6 Mohs.

224

CEMENT - is a fired mixture of clay and limestone. This is ground into a powder so that it may be mixed with water to further react. Some properties are listed under concrete in the data section.

CEMENTED CARBIDES - is a composite of refractory ceramics in a matrix of high ductility metal. Usually they are used as a cutting tool.

CERAMIC - is an inorganic non-metallic material. Sometimes this term is restricted to crystalline materials of the above description.

CERARGYRITE - is the mineral form of AgCl. See the data section for the properties.

CERIA - is an alternate name for cerium oxide.

CERIUM OXIDE - CeO_2. See the data section for the properties.

CERMETS - is a composite of ceramic and metal. This is a general term for a material with a ceramic content between 30% and 70%. Dispersion-hardened metal is the term for a composite with a low ceramic content.

CERUSSITE - is the mineral form of $PbCO_3$. It has an index of refraction of 1.8-2.1, a specific gravity of 6.6, a decomposition temperature of 315° C and a solubility in water of 0.0001 g/100ml.

CERVANTITE - is the mineral form of antimony tetraoxide. It has an index of refraction of 2.0, a specific gravity of 5.82 and a melting point of 930° C.

CESIUM CHLORIDE - CsCl. It has a cubic structure, an index of refraction of 1.64, a specific gravity of 3.99, a melting temperature of 645° C, a boiling temperature of 1290° C, a solubility of 162 g/100 cc in cold water and a solubility of 260 g/100cc in hot water. The structure is shown in Figure 1-5.

CESIUM IODIDE - CsI. See the data section for the properties.

CHALCANTHITE - is the mineral form of $CuSO_4 \cdot 5H_2O$. It has a triclinic structure, an index of refraction of 1.5, a specific gravity of 2.3, a solubility in cold water of 31.6 g/100ml and in hot water of 203.3 g/100ml.

CHALCEDONY - is a mineral form of silica. See the data section for properties.

CHALK - is a less dense mineral form of calcium carbonate. It has a specific gravity of 1.9.

225

CHAOITE - is a form of carbon with a hexagonal structure found in nature only where meteorites have impacted the earth.

CHEMICAL VAPOR DEPOSITION - is a process of creating a ceramic material which includes a chemical reaction. This process is usually done near room temperature without a temperature gradient. The resulting material is very pure, has low residual stress and generally displays a cone or plate macrostructure. It is used to produce complex finished shapes and coatings.

CHERT - is a naturally occurring very finely grained quartz with many inclusions of water. The color is opaque white to light brown.

CHESSYLITE - is an alternate term for azurite.

CHEVREUL'S SALT - $Cu_2SO_3 \cdot CuSO_3 \cdot 2H_2O$. It is a red pigment with a specific gravity of 3.6 and no solubility in water.

CHINA CLAY - is an alternate term for kaolin.

CHLORITE - is a clay composed of alternating layers of talc and brucite. Some of the magnesium and silicon may be replaced by aluminum in this structure.

CHROME RED - $PbCrO_4 \cdot PbO$. It is a red pigment with a specific gravity of 6.6 and no solubility in water.

CHROME YELLOW - is an alternate term for crocoite.

CHROMIA - is an alternate term for chromium oxide.

CHROMIC OXIDE - is an alternate term for chromium oxide.

CHROMITE - is an alternate term for chromium oxide.

CHROMIUM BROIDE - $CrBr$. See data section for properties.

CHROMIUM OXIDE - Cr_2O_3. See data section for properties.

CHRYSOBERYL - is the mineral form of beryllium aluminate. It has a rhombic structure, an index of refraction of 1.75, a specific gravity of 3.76 and a melting temperature of 1870° C.

CHRYSOTILE - is a type of asbestos. See the data section under asbestos for the properties.

CIMOLITE - $2Al_2O_3 \cdot 9SiO_2 \cdot 6H_2O$

CLAUDETITE - is the mineral form of arsenic trioxide. See the data section for the properties.

CLAUSTHALITE - is the mineral form of PbSe. It has a cubic structure, a specific gravity of 8.1, a melting temperature of 1065° C and no solubility in water.

CLAY - is formed by the decomposition of feldspar. If the deposit occurs at the site of the decomposition of the rock than it is a primary clay and will most likely fire to a white color and be brittle and refractory. Clay that has moved from the original site and mixed with other things is secondary clay which is more variable in nature. The most common components of clay are kaolinite, illite and quartz.

CLAYITE - $Al_2O_3 \cdot 2SiO_2 \cdot 2H_2O$

CLINOENSTATITE - is the mineral form of $MgSiO_3$. It has a monoclinic structure, a specific gravity of 3.2, a decomposition temperature of 1557° C, a coefficient of thermal expansion of $7.8X10^{-6}$/C over 100° to 200° C, and an insolubility in water. It can be converted from orthoenstatite at 1140° C.

CLINKER - see Portland cement.

COBALT CARBONATE - $CoCO_3$. The mineral form is spherocobaltite. It is used as a blue coloring agent in glazes.

COBALTITE - is the mineral form of CoAsS. It has a specific gravity of 6.2-6.3, and it decomposes on heating.

COLEMANITE - is the mineral form of $2CaO \cdot 3B_2O_3 \cdot 5H_2O$. It is insoluble in water and is used in glazes, especially ones containing lead.

COLLYRITE - $2Al_2O_3 \cdot SiO_2 \cdot 9H_2O$

CONCRETE - is a composite ceramic in which aggregate is bound with cement. See the data section for the properties.

CONE - is a contraction for pyrometric cone equivalent.

CONTAINER GLASS - has a common composition of 72.1% silica, 1.8% alumina, 0.1% ferric oxide, 5.6% calcium oxide, 4.2% magnesia, 0.3% barium oxide and 15.6% sodium oxide.

COPPER BROMIDE - Cu_2Br_2. See the data section for the properties.

COPPER CARBONATE - Cu_2CO_3. It has a specific gravity of 4.4, and it is insoluble in water. Related mineral forms are malachite azurite and chessylite. Depending on the conditions in a glaze, it yields green or red.

COPPER CHLORIDE - Cu_2Cl_2. The mineral form is nantokite.
COPPER IODITE - CuI. The mineral form is marshite.

COPPER SULFATE - $CuSO_4$. The mineral form is hydrocyanite.

COQUIMBITE - is the mineral form of $Fe_2(SO_4)_3 \cdot 9H_2O$. It has a rhombic structure, an index of refraction of 1.6, a specific gravity of 2.1, and a solubility in water of 440 g/l00ml.

CORDIERITE - $2MgO \cdot 2Al_2O_3 \cdot 5SiO_2$. See the data section for the properties.

CORUNDUM - is the term for the alpha phase of aluminum oxide as well as the name of the mineral form. See the data section for properties.

COTUNITE - is the mineral form of $PbCl_2$. It has a rhombic structure, an index of refraction of 2.2, a specific gravity of 5.9, a melting temperature of 501^0 C, a boiling temperature of 950^0 C, a solubility in cold water of 0.99 g/100ml and in hot water of 3.34 g/100ml.

COVELLITE - is the mineral form of CuS. It has a rhombic structure, a specific gravity of 5.6, a melting temperature of 1100^0 C and a slight solubility in water.

CRISTOBALITE - is a mineral form of silicon dioxide which is chemically similar to quartz, lechatelierite and tridymite. See the data section for the properties.

CROCIDULITE - is a form of asbestos. See the data section under asbestos for the properties.

CROCUS MARTIS - is a glaze that produces a mottled brown and yellow affect on stoneware.

CROWN GLASS - has a common composition of 72.2% silica, 5.9% boric oxide, 0.2% arsenic oxide, 2.1% calcium oxide, 0.1% magnesia, 13.9% potassium oxide, 5.2% sodium oxide, 0.1% antimony oxide and 0.1% sulfur trioxide.

CRYOLITE - is the mineral form of $3NaF \cdot AlF_3$. It has a specific gravity of 2.98, a hardness of 2.5 Mohs, a solubility in water of 0.04 g/100ml. at 25^0 C, a melting temperature of 1020° C. Cryolite is used as a flux and an opacifer for glazes. A major disadvantage to its use in glazes is its formation of dangerous fluorine fumes.

CRYSTAL - is a grade of glass with a high lead oxide content. It is usually cut or engraved to display its high index of refraction. Originally this was made from flint but proved to be unstable. Lead oxide was added to aid stability and eventually completely replaced the flint. A common composition for crystal is 67.4% silica, 1.7% alumina, 3.9% zincite, 0.4% calcium oxide, 10.7% lead oxide, 0.1% potassium oxide and 15.1% sodium oxide.

CUPRITE - is the mineral form of Cu_2O. It has a cubic structure, an index of refraction of 2.7., a specific gravity of 6.0, a melting temperature of 1235° C and an insolubility in water.

CYANITE - is a mineral form of aluminum silicate. See the data section for the properties.

D-DRYING - brings concrete into equilibrium at a water vapor pressure of 0.0005 mm. of Hg.

DEFLOCCULANT - is used as an additive to a particle slurry to reduce the tendency of the particles to form a solid mass.

DEVITRIFICATION - is the conversion of a glass from the glassy state to the crystalline state. This is usually an undesirable consequence of processing.

DIAMOND - is a mineral form of carbon. See the data section for properties.

DIASPORE - is a mineral from of aluminum oxide monohydrate. It has a rhombic structure and a specific gravity of 3.3-3.5 The low temperature form of this mineral is bauxite.

DIATOMACEOUS EARTH - see diatomite.

DIATOMITE - is the comprised of the skeletons of diatons in which the main component is amorphous silica. It is highly porous and has a very low thermal conductivity.

DICKITE - $Al_2O_3 \cdot 2SiO_2 \cdot 2H_2O$. It is a member of the kaolinite group of clays and has a thin hexagonal plate structure.

DIFFUSION BONDING - is a process wherein two dissimilar surfaces are brought together and subjected to high pressure and temperature. If these conditions are held for a long enough time the atoms diffuse into each other creating a bond. Usually this is done between a ceramic and a metal.

DILITHIUM SODIUM PHOSPHATE - $2Li_2O\cdot Na_2O\cdot P_2O_5$. It has a rhombic structure, a hardness of 4-5 Mohs and a specific gravity of 3.4-3.6

DIOPSIDE - is the mineral form of calcium magnesium metasilicate. It has a monoclinic structure, an index of refraction of 1.7, a specific gravity of 3.3, a melting temperature of 1390° C and no solubility in water.

DISCONTINUOUS GRAIN GROWTH - is a process wherein the grain boundaries break away from included pores. As a result the pore is included within the larger grain that continues to grow. Pores inside a grain take a much longer time at elevated temperatures to be eliminated than pores at or near the grain boundary.

DISPERSION-HARDENED METAL - refers specifically to a metal with a very small sized particle ceramic distributed throughout it. The ceramic phase will not exceed 15%.
DOLOMITE - is the mineral form of calcium magnesium carbonate. It has a trigonal structure, a specific gravity of 2.9, an index of refraction of 1.5-1.7, a solubility in water of 0.03 g/100ml., a hardness of 3.5-4 Mohs and a decomposition temperature of 730° C.

DOMEYKITE - is the mineral form of Cu_3As. It has a hexagonal structure, a specific gravity of 8.0 and a melting temperature of 830° C.

DOPING - is the addition of very small quantities of a material not found in the bulk of the ceramic. Usually doping is done by movement of the new atoms into interstitial sites.

DRYING (clay) - is done by slowly heating the slip such that the free water evaporates without creating steam. Further heating is used to remove the chemically combined water. This begins at about 300° C and is completed about 500° C.

DUMORTIERITE - is the mineral form of $8Al_2O_3\cdot 6SiO_2\cdot B_2O_3\cdot H_2O$. It has a specific gravity of 3.2-3.3 and a hardness of 7 Mohs. Dumortierite breaks down to mullite and silica at 1230° C.

EMORY - $Al_2O_3\cdot Fe_3O_4$. See data section for properties.

ENAMEL - is a glaze that is fired below 800° C and usually alters the color of the final product.

ENSTATITE - is an alternate term for clinoenstatite.

EPSOM SALT - is an alternate term for epsomite.

230

EPSOMITE - is the mineral form of $MgSO_4 \cdot 7H_2O$. It has a rhombic or monoclinic structure, an index of refraction of 1.4-1.5, a specific gravity of 1.7, a solubility in water of 71 g/100ml. at 20° C and 91 g/100ml. at 40° C. Epsomite is used as a flocculant to stiffen casting slips.

ERDMANNIS SALT - $NH_4(Co(NH_3)_2(NO_2)_4)$. It has a rhombic structure, an index of refraction of 1.8 and a specific gravity of 1.9.

ERIOCHALEITE - is the mineral form of $CuCl_2 \cdot 2H_2O$. It has a rhombic structure, an index of refraction of 1.7, a specific gravity of 2.5, a solubility in cold water of 110 g/100ml and in hot water of 192 g/100ml.

ERYTHROSIDERITE - is the mineral form of $2KCl \cdot FeCl_3 \cdot H_2O$. It has an orthorhombic structure and a specific gravity of 2.4

ETCHING - is the reaction of soluble radicals in a solution with the components of a glass. An example is the attack of oxygen in a silicon dioxide glass by sodium. The sodium first breaks the glass backbone bond and then removes the oxygen from the glass structure.

ETTRINGITE - is $3CaS \cdot 31H_2O$. It is an intermediate product of calcium sulfide and Ca_3AlO in Portland cement.

EUCLASE - is the mineral form of $Be_2Al_2(SiO_4)_2 \cdot (OH)_2$. It has a monoclinic structure, an index of refraction of 1.66 and a specific gravity of 3.1.

EUCRYPTITE - is the mineral form of $Li_2O \cdot Al_2O_3 \cdot 2SiO_2$. It has a hexagonal structure, a specific gravity of 2.67 in the alpha phase and 2.35 in the beta phase, the conversion from the alpha to beta phase takes place at 970° C and the melting temperature of the beta phase is about 1400° C.

EULYTITE - is the mineral form of bismuth silicate. It has a cubic structure, an index of refraction of 2.1 and a specific gravity of 6.1.

FAIENCE - is a glazed, fired ceramic which is usually earthenware.

FAVAS - is a mineral that contains about 85% zirconium oxide.

FAYALITE - is the mineral form of Fe_2SiO_4. It has a rhombic structure, a specific gravity of 4.3, a melting temperature of 1503° C and no solubility in water.

FELDSPAR - is a mixture of potash, soda and lime. It serves as a source of alumina and as a flux in glass making. The nominal composition is $K_2O \cdot Al_2O_3 \cdot 6SiO_2$.

FERBERITE - is the mineral form of $FeWO_4$. It has a tetragonal structure, and a specific gravity of 6.6.

FERRIC OXIDE - Fe_2O_3. See the data section for the properties.

FERRITE - is an alternate term for hexaferrite.

FERROCEMENT - is a modified form of reinforced concrete. It usually contains Portland cement without an aggregate which is continuously reinforced by small diameter wire mesh.

FERROUS OXIDE - FeO. The mineral form is wuestite.

FIBER GLASS - is a textile made of glass. The common composition of it is 54% silica, 14.0% alumina, 17.5% calcium oxide and 4.5% magnesium oxide.

FICTIVE TEMPERATURE - is only used relative to inorganic glasses and is defined as the stable temperature for a given volume. This is not the true equilibrium temperature as that would only be applicable to a crystal.

FIRE CLAY - is refractory clay containing mainly livesite, hydrous mica and quartz. It is usually used for inexpensive crucibles because its low strength and porous texture limit it.

FIRING (clay) - is a process which is considered to commence at 573° C when the free silica changes phase from alpha to beta quartz. Around 1100° C the quartz begins to crystalize into cristobalite. Cristobalite is a stable phase and does not revert as beta quartz does on cooling. Generally the higher the final temperature reached during firing the stronger the clay and the more it will shrink.

FLINT - is the mineral form of $SiO_2 \cdot xH_2O$. It is usually gray due to included carbonaceous material.

FLINT CLAY - is an unusual type of clay containing a high amount of silica.

FLINT GLASS - is an alternate term for crystal.

FLUELLITE - is the mineral form of $AlF_3 \cdot H_2O$. It has a rhombic structure, a specific gravity of 2.17 and an index of refraction of 1.5.

FLUORITE - is the mineral form of calcium fluoride. See the data section for the properties.

FLUORSPAR - is an alternate term for fluorite.

FLUX - is an additive which interacts with the ceramic and causes some or all of the components to melt. In brick making the most common flux is calcium. This term can also be used to mean the pigment binder in a glaze.

FLY ASH - is also known as fuel ash. It is the residue from powdered coal combustion. The composition is mainly glass phase aluminum silicates, and it finds use in the concrete industry.

FORSTERITE - is the mineral form of $2MgO \cdot SiO_2$. See the data section for the properties.

FREE SILICA - is silica which is not chemically bound to clay or other ceramics in a material. It is thus free to change its phase with prevailing conditions without modification due to other bonds.

FRENKEL DEFECT - is a combination of vacancies and atoms in neighboring interstitial sites that constitute a stoiciometric group. Since the vacancy or vacancies associated with this group have an effective charge of the atom or atoms in the interstitial sites they will remain in the vicinity until extra energy is added to the ceramic, usually thermal energy, so that they can break away.

FRIT - is a glassy material that has not been fully processed to the final glaze form and contains some of the impurities of its natural state.

FSZ - is an acronym for fully stabilized zirconia.

GAHNITE - is the mineral form of $ZnAl_2O_4$. It has a cubic structure, an index of refraction of 1.8, a specific gravity of 4.6 and no solubility in water.

GALENA - is the mineral form of PbS. It has a cubic structure, an index of refraction of 3.9, a specific gravity of 7.5, a melting temperature of 1114° C, and is slightly soluble in water.

GARNET - is the mineral form of $Al_2O_3 \cdot 3FeO \cdot 3SiO_2$. See the data section for the properties. This is also a general term for a wide range of minerals with alumina and silica as the base.

GIBBSITE - is the mineral form of aluminum oxide trihydrate. It has a layered structure and a specific gravity of 2.4.

GLASS-CERAMIC - is a composite material produced by first forming a glass into the shape desired and then heat treating to produced some crystalline phase within the glass. For the most part these materials display the properties of the glass phase because the glass is the continuous phase up to the point of very high crystallinity.

GLASS TYPE NUMBERING - is a system wherein glasses are numbered by type with a six digit number separated by a colon in the center. The first three digits are the digits following the decimal point of the refractive index. Because all glasses have a 1 before the decimal place it was not considered necessary to include this. The last three digits are the ABBE number without the decimal point. ABBE numbers are always two digits on the left of the decimal place thus the decimal point was also considered unnecessary.

GLAZE - is a glassy film used as a coating. Glazes add shine and seal surface pores. They are produced by spraying or painting on a slurry and then firing it. Almost all glazes are silica based.

GLAZE FIT - is the matching of the thermal expansion and contraction of a glaze and the ceramic under to reduce stress during thermal cycling.

GOSLARITE - is the mineral form of $ZnSO_4 \cdot 7H_2O$. It has a rhombic structure, an index of refraction of 1.5, a specific gravity of 2.0, a melting temperature of 100^0 C, a solubility in cold water of 97 g/100ml and in hot water of 664 g/l0Oml.

GPSSN - is an acronym for gas-pressure-sintered silicon nitride.

GRAIN GROWTH - is the movement of atoms via diffusion from a smaller grain to a larger one. The smaller grain has larger surface area and the extra energy associated with it is the driving force. Eventually small grains will disappear.

GRANULATED SLAG - is the glassy state slag from an iron producing blast furnace. It contains the chemicals necessary to make clinker but the proportions must be adjusted to result in effective cement.

GRAPHITE - is the hexagonal form of carbon and the stable phase at ambient. Natural graphite is heat treated above 2000^0 C to form an artificial form which is even more stable.

GRAPHITE FIBER - is a fiber which is 98% or greater carbon.

GREEN CLAY (greenware) - is clay that is firm but has yet to be fired.

GREENOCKITE - is the mineral form of cadmium sulfide. See the data section for the properties.

GROG - is fired clay that has been ground into small particles and mixed back into a clay that is to be fired. It serves as an additive that reduces shrinkage during firing. It also increases the overall particle size which increases the capillaries in the clay. Larger capillaries facilitate drying.

GRUENERITE - is the mineral form of $FeSiO_3$. It has a rhombic structure, a specific gravity of 3.5, an index of refraction of 1.7 and a melting temperature of 1146° C.

GUANAJUATITE - is the mineral form of Bi_2Se_3. It has a rhombic structure, a specific gravity of 6.8 and a melting temperature of 710° C.

GYPSUM - is the mineral form of $CaSO_4 \cdot 2H_2O$. It has a monoclinic structure, an index of refraction of 1.5, a specific gravity of 2.3 and a solubility in water of 0.2 g/100ml.

HAFNIUM CARBIDE - HfC. See the data section for the properties.

HAIDINGERITE - is the mineral form of $2CaO \cdot As_2O_5 \cdot 3H_2O$. It has a rhombic structure, an index of refraction of 1.6 and a specific gravity of 3.0

HALITE - is the mineral form of sodium chloride. See the data section for the properties.

HALLOYSITE - $Al_2O_3 \cdot 2SiO_2 \cdot xH_2O$. It is a member of the kaolite group of clays. Halloysite exists in both a low temperature and a high temperature version and has a thin hexagonal structure.

HAMBERGITE - is the mineral form of $Be_2(OH)BO_3$. It has a rhombic structure, an index of refraction of 1.6, and a specific gravity of 2.35

HATCHETT'S BROWN - $CuFe(CN)_6 \cdot xH_2O$. It is a redish brown pigment which is insoluble in water.

HAUERITE - is the mineral form of MnS_2. It has a cubic structure, a specific gravity of 3.5 and no solubility in water.

HAUSMANNITE - is the mineral form of Mn_3O_4. It has a tetragonal structure, a specific gravity of 4.9, a melting temperature of 1564° C and no solubility in water.

HEAVY SPAR - is an alternate term for barytes.

HEMATITE - is a mineral form of ferric oxide. See the data section for the properties.

HEMIHYDRATE - $2CaSO_4 \cdot H_2O$. It has a solubility in water of 0.3 g/100ml and is produced by removing some of the bound water in gypson at 170° C.

HEMIMORPHITE - is the mineral form of $2ZnO \cdot SiO_2 \cdot H_2O$. It has a rhombic structure, an index of refraction of 1.6, a specific gravity of 3.5 and no solubility in water.

HESSITE - is the mineral form of Ag_2Te, it has a cubic structure, a specific gravity of 8.5, a melting temperature of 955° C and no solubility in water.

HEXAFERRITE - is a class of ceramics generally used as permanent magnets. See the data section for the properties.

HIERATITE - is the mineral form of K_2SiF_6. It has a cubic or hexagonal structure, an index of refraction of 1.4, a specific gravity of 3.1 for the hexagonal form and 2.7 for the cubic form. The solubility in water for the cubic form is 7 g/100ml., the solubility in cold water for the hexagonal form is 0.12 g/100ml and in hot water for the hexagonal form is 0.95 g/l00ml.

HIGH ALUMINA CEMENT - is this is composed of 36%-40% alumina, 36%-40% calcium oxide, 16% iron oxides and 5% silica. Its resistance to acids is very good. However, it is attacked by the alkaline components of Portland cement and is therefore incompatible with it. High alumina cement suffers from low strength and takes a very long time to set. Its main use is as a refractory because it can withstand temperatures up to 1600° C.

HIPSN - is an acronym for hot isostatically pressed silicon nitride.

HOERNESITE - is the mineral form of $Mg_3(AsO_4)_2 \cdot 8H_2O$. It has a monoclinic structure, and a specific gravity of 2.6.

HPSN - is an acronym for hot pressed silicon nitride.

HYALOPHANE - is a mixture of orthoclase and celsian.

HYDRARGILLITA - is the mineral form of aluminum oxide trihydrate. It has a specific gravity of 2.4

HYDRAULIC CEMENT - falls under the general definition of cement.

HYDROCYANITE - is the mineral form of copper sulfate. It has a rhombic structure, an index of refraction of 1.7, a specific gravity of 3.6, a solubility in cold water of 14 g/100ml and in hot water of 75 g/100ml. It decomposes to form tenorite at 650° C.

HYDROMANGNESITE - is the mineral form of $3MgCO_3 \cdot Mg(OH)_2 \cdot 3H_2O$. It has a rhombic structure, an index of refraction of 1.5, a specific gravity of 2.2, a solubility in cold water of 0.04 g/100ml and in hot water of 0.01 g/100ml.

ILLITE - is a group of clays which do not absorb water. They are held together by a layer of potassium. The general formula is $(OH)_4K_Y(Al_4 \cdot Fe_4 \cdot Mn_4 \cdot Mg_6)Si_{8-Y}Al_YO$ where Y varies from 1 to 1.5

ILMENITE - is a mineral form of $FeO \cdot TiO_2$. It has a specific gravity of 4.8, and a hardness of 5-6 Mohs. Ilmenite produces brown speckles in glazes.

ILMENORUTILE - is a mineral which combines rutile, niobium and iron.

INCLUSION PIGMENTS - are ceramic pigments which are encapsulated by zircon. The zircon protects them from being dissolved by the glaze and thus altering their coloration. This technique allows for both a wider range of colors and temperatures.

IODYRITE - is the mineral form of AgI. See the data section for the properties.

IRON CARBONATE - $FeCO_3$. The mineral form is siderite.

IRON DISULFIDE - FeS_2. See the data section for the properties.

JADEITE - is the mineral form of $Na_2O \cdot Al_2O_3 \cdot 4SiO_2$. It has a monoclinic structure, a specific gravity of 3.3, a melting temperature of 1000° to 1060° C, a color of green and no solubility in water.

KAINITE - is the mineral form of $KCl \cdot MgSO_4 \cdot 3H_2O$. It has a monoclinic structure, a specific gravity of 21 and a solubility in water of 80 g/100ml.

KALINITE - is the mineral form of $KAl(SO_4)_2 \cdot 12H_2O$. It has a cubic, hexagonal or monoclinic structure, an index of refraction of 1.4-1.5, a specific gravity of 1.8 and a solubility in water of 11 g/100ml.

KALIOPHITE - is the mineral form of $KAlSiO_4$. It has a hexagonal structure which converts to rhombic at 1540° C, an index of refraction of 1.5-1.6, a specific gravity of 2.5 and a melting temperature of 1800° C.

KAOLIN - is a clay containing mainly kaolinite after washing. Millions of tons each year are used in the production of paper finish. It is also used in the plastics and paint industries. The terms clay and china clay are often used to indicate kaolin.

KAOLINITE - is a clay mineral with the composition $Al_2O_3 \cdot 2SiO_2 \cdot 2H_2O$. It has a thin plate like crystalline structure bound together by hydrogen bonds. The plates

237

are composed of alternating layers of silica and gibbsite. Its melting temperature is 1770° C.

KERNITE - is the mineral form of $Na_2O \cdot 2B_2O_3 \cdot 4H_2O$. It is used as a flux in glazes.

KIESELGUHR - is an alternate term for diatomite.

KIESERITE - is the mineral form of $MgSO_4 \cdot H_2O$. It has a monoclinic structure, an index of refraction of 1.5-1.6, a specific gravity of 2.4 and a solubility in hot water of 68 g/l00ml.

KNOPITE - is the mineral form of $TiO_2 \cdot CeO$.

KOCHITE - is the mineral form of $2Al_2O_3 \cdot 3SiO_2 \cdot 5H_2O$.

KOETTIGITE - is the mineral form of $Zn_3(AsO_4)_2 \cdot 8H_2O$. It has a monoclinic structure, a specific gravity of 3.3, and index of refraction of 1.7 and no solubility in water.

KRAUSITE - is the mineral form of $K_2SO_4 \cdot Fe(SO_4)_3 \cdot 24H_2O$. It has a monoclinic structure, an index of refraction of 1.5, a specific gravity of 1.8, a melting temperature of 28° C and a decomposition temperature of 33° C.

KRENNERITE - is the mineral form of $AuTe_2$. It has a specific gravity of 8.2-9.3, a decomposition temperature of 472° C and no solubility in water.

KYANITE - is an alternate spelling for cyanite.

LABRADORITE - is a feldspar intermediate in composition between albite and anorthite .

LAITANCE - is a mixture of cement, fine sand and water which forms when the course aggregate of concrete settles before setting.

LANARKITE - is the mineral form of $PbSO_4 \cdot PbO$. It has a monoclinic structure, an index of refraction of 2.0, a specific gravity of 6.9, a melting temperature of 977° C and a solubility in water of 0.004 g/100ml.

LANGBELINITE - is the mineral form of $K_2SO_4 \cdot 2MgSO_4$, It has a tetragonal structure, a specific gravity of 2.8 and a melting temperature of 927° C.

LANSFORDITE - is the mineral form of $MgCO_3 \cdot 5H_2O$. It has a monoclinic structure, an index of refraction of 1.5, a specific gravity of 1.7, a solubility in cold water of 0.18 g/100ml and in hot water of 0.38 g/100ml.

LAURIONITE - is the mineral form of $PbCl_2 \cdot Pb(OH)_2$. It has a rhombic structure, a specific gravity of 6.2 and a decomposition temperature of 142° C.

LAURITE - is the mineral form of RuS_2. It has a cubic structure, a specific gravity of 7.0, a decomposition temperature of 1000° C and no solubility in water.

LAUTARITE - is the mineral form of calcium iodate. It has a monoclinic structure, a specific gravity of 4.5, a decomposition temperature of 540° C, a solubility in cold water of .20 g/100ml and in hot water of 0.67 g/100ml.

LAWRENCITE - is the mineral form of $FeCl_2$. It has a hexagonal structure, an index of refraction of 1.6, a specific gravity of 3.16, a melting temperature of 670° to 674° C, a solubility in cold water of 65 g/100ml and in hot water of 106 g/100ml.

LEACHING - is a reaction between atoms located in the interstitial sites of a glass and a liquid with soluble reactants. The main structure of the glass is unaffected by this.

LEAD CRYSTAL - is an alternate term for crystal.

LEAD OXIDE - PbO. The mineral forms are massicot and litharge. See the data section for the properties.

LECHATELIERITE - is a mineral form of silicon dioxide chemically the same as cristobalite, trydymite and quartz. See the data section for the properties.

LEONITE - is the mineral form of $K_2SO_4 \cdot MgSO_4 \cdot 4H_2O$. It has a monoclinic structure, an index of refraction of 1.5, a specific gravity of 2.2 and a high solubility in water.

LEPIDOLITE - is a naturally occurring frit of lithium, silica and alumina. It is usually $2Li_2O \cdot K_2O \cdot 2Al_2O_3 \cdot 6SiO_2 \cdot 2F_2 \cdot H_2O$. Lepidolite has a monoclinic structure, a specific gravity of 2.8-2.9 and a hardness of 3 Mohs.

LEUCITE - is the mineral form of $KAlSi_2O_6$. It has an index of refraction of 1.5, a specific gravity of 2.5, a melting temperature of 1686° C and no solubility in water.

LEVERRIERITE - is the mineral form of $Al_2O_3 \cdot 3SiO_2 \cdot 3H_2O$.

LIME - is a common term for calcium oxide.

LIME SPAR - is the common term for anorthite.

LIMESTONE - is a mineral form of calcium carbonate. It has a specific gravity of 2.3-2.6 and a hardness of 3-4 Mohs.

LINNEITE - is the mineral form of Co_3S_4. It has a cubic structure, a specific gravity of 4.9 and a decomposition temperature of 480° C.

LITHARGE - is the mineral form of lead oxide. See the data section for properties.

LITHIA - is an alternate term for lithium oxide.

LITHIUM ALUMINATE - $LiAlO_2$. It has a melting temperature of 1625° C., and it is used as a source of coloration.

LITHIUM CARBONATE - Li_2CO_3. It has a specific gravity of 2.1, a melting temperature of 618° C and a solubility in water of 1.54 g/100ml at 0° C, 1.33 g/100ml at 20° C and 0.72 g/100ml at 100° C. It is used as a frit in glazes and as a defloccu-lant in clay.

LITHIUM FLUORIDE - LiF. See the data section for the properties.

LITHIUM OXIDE - Li_2O. It has a specific gravity of 2, a melting temperature above 1700° C and a solubility in water of 6.67 g/100ml at 0° C and 10.02 g/100ml at 100° C.

LITHIUM SILICATE - Li_2SiO_3. It has a specific gravity of 2.52, a coefficient of thermal expansion of $11X10^{-6}$/C for 20° to 600° C, and a melting temperature of 1201° C. It is used as a frit in glazes.

LITHIUM TITANATE - Li_2TiO_3. It is used as a flux.

LITHIUM ZIRCONATE - Li_2ZrO_3. It is used as a flux.

LITHIUM ZIRCONIUM-SILICATE - $2Li_2O \cdot ZrO_2 \cdot SiO_2$. It is used as a mill addi-tion to tile glazes for enhanced smoothness of finish.

LIVESITE - is a member of the kaolinite group of clays. It has a thin hexagonal structure.

LOESS - is a windborne sedimentary rock form of clay. It usually contains quartz, feldspar and calcium carbide. It is also called adobe and is mainly used in brick making.

LONG GLASS - is glass with a small variation of viscosity as a function of temperature. As a result there is a long workable region.

LONSDALEITE - is a form of carbon which is hexagonal in crystalline structure but still a kind of diamond. This alternate to the normal cubic structure of diamond was produced during diamond synthesization.

LOPARITE - is the mineral perowskite with the addition of niobium.

MAGNANITE - is the mineral form of MnO(OH). It has a rhombic structure, an index of refraction of 2.2, a specific gravity of 4.2-4.4 and no solubility in water.

MAGNESIA - is an alternate term for magnesium oxide.

MAGNESITE - is the mineral form of magnesium carbonate. It has a specific gravity of 3, a trigonal structure and a hardness of 3.5-4.5 Mohs. It decomposes at 350° C and its solubility in water is 0.016 g/100ml.

MAGNESIUM ALUMINATE - $MgAl_2O_4$. The mineral form is spinel. See the data section for properties.

MAGNESIUM CARBONATE - $MgCO_3$. The mineral form is magnesite.

MAGNESIUM OXIDE - MgO. The mineral form is periclase. See the data section for properties.

MAGNETITE - is a mineral form of ferric oxide. See the data section for the properties.

MAIOLICA - is earthenware glazed with white tin. It is often decorated.

MALACHITE - is the mineral form of $CuCo_3 \cdot Cu(OH)_2$. It has a monoclinic structure, an index of refraction of 1.7-1.9, a specific gravity of 4.0, a decomposition temperature of 200° C and no solubility in water.

MANGANESE - is used in glass making because it produces a pink cast which balances the normal green shade in glass due to the presence of iron impurities.

MANGANESE DIOXIDE - MgO_2. The mineral form is pyrolusite. This is the final form that the manganese takes in glass and usually constitutes less than 5% of a glass mixture. It produces a wide range of colors in glazes.

MANGANOSITE - is the mineral form of MnO. It has a cubic structure, an index of

241

refraction of 2.2, a specific gravity of 5.4 and no solubility in water.

MARBLE - is a mineral form of calcium carbonate with a specific gravity of 2.7 and a hardness of 3-4 Mohs.

MARCASITE - is the mineral form of iron disulfide. See the data section for the properties.

MARLS - is a mixture of clay and chalk.

MARSHITE - is the mineral form of copper iodide. See the data section for the properties.

MARTENSITIC TRANSFORMATION - is a change from one crystalline type to another via shear. Since diffusion is not involved this reaction is essentially independent of time. The crystalline type changes due to the transformation, thus a phase change occurs even though no compositional change has taken place.

MASSICOT - is the mineral form of lead oxide. See the data section for properties.

MATLOCKITE - is the mineral form of PbFCl. It has a tetragonal structure, an index of refraction of 2.1, a specific gravity of 7.1, a melting temperature of 601° C, a solubility in cold water of 0.04 g/100ml and in hot water of 0.11 g/lOOml.

MELANTERITE - is the mineral form of $FeSO_4 \cdot 7H_2O$. It has a monoclinic structure, an index of refraction of 1.5, a specific gravity of 1.9, a solubility in cold water of 16 g/100ml and in hot water of 49 g/100ml.

MENDIPITE - is the mineral form of $PbCl_2 \cdot 2PbO$. It has a rhombic structure, an index of refraction of 2.2-2.3, a specific gravity of 7.1, a melting temperature of 693° C and no solubility in water.

MERCALLITE - is the mineral form of $KHSO_4$. It has a rhombic structure, an index of refraction of 1.5, a specific gravity of 2.3, a melting temperature of 214° C, a solubility in cold water of 36 g/100ml and in hot water of 122 g/lOOml.

MERVINITE - is the mineral form of calcium magnesium orthosilicate. It has a monoclinic structure, an index of refraction of 1.7, and a specific gravity of 3.2

METAL-INFILTRATED CERAMIC - is a composite material made with a sintered ceramic into which a metal is diffused. This usually requires the aid of a metal oxide to reduce the surface tension of the metal. The final structure is one in which both the metal and the ceramic are continuous.

METALLIZATION - is the application of a thin film of metal on glass without soft-

242

ening the glass.

MICA - is a mineral with the composition $(0.3)K_2O \cdot 3Al_2O_3 \cdot 6SiO_2$. It usually decomposes to hydrous mica or illite. See potassium aluminosilicate for the approximate properties.

MICA (hydrous) - is a mineral with the composition
$(0.3)K_2O \cdot 3Al_2O_3 \cdot 6S iO_2 \cdot (4.5)H_2O$

MICROCLIME - is a mineral form of potassium aluminosilicate. See the data section for the properties.

MILLERITE - is the mineral form of NiS. It has a specific gravity of 5.3-5.7, a trigonal structure, a melting temperature of 797° C and a solubility in water of 0.0004 g/l00ml.

MINERAL - is naturally occurring ceramic material.

MISENITE - is an alternate name for mercallite.

MOLYBDENITE - is the mineral form of MoS_2. It has a hexagonal structure, a specific gravity of 4.8, a melting temperature of 1185° C and no solubility in water.

MOLYBDENUM CARBIDE - Mo_2C. It has a hexagonal structure, a specific gravity of 8.9, a melting temperature of 2687° C, a modulus of elasticity of 30,000,000 psi and no solubility in water.

MOLYBDENUM SILICIDE - $MoSi_2$. See the data section for the properties.

MOLYSITE - is the mineral form of FeCl. It has a hexagonal structure, a specific gravity of 2.9, a melting temperature of 306° C, a solubility in cold water of 74 g/100ml and in hot water of 536 g/100ml.

MONAZITE - is the mineral form of $CePO_4$. It has a monoclinic or rhombic structure, an index of refraction of 1.8, a specific gravity of 5.2 and no solubility in water.

MONIMOLITE - is the mineral form of $Pb_3(SbO_4)_2$. It has a specific gravity of 6.6 and no solubility in water.

MONTROYDITE - is the mineral form of HgO. It has a rhombic structure, an index of refraction of 2.4-2.7, a specific gravity of 11.1, a decomposition temperature of 500° C, a solubility in cold water of 0.005 g/100ml and in hot water of 0.04 g/100ml.

243

MONTMORILLONITE - is the mineral form of $Al_2O_3 \cdot 4SiO_2 \cdot xH_2O$.

MORTAR - is a mixture of sand and cement.

MOSSITE - is a combination of rutile and niobium.

MULLITE - is the mineral form of aluminum silicate, $3Al_2O_3 \cdot 2SiO_2$.. See the data section for the properties.

MUSCOVITE - is the mineral form of $K_2O \cdot 3Al_2O_3 \cdot 6SiO_2 \cdot 2H_2O$. It has a mono-clinic structure, an index of refraction of 1.6, a specific gravity of 2.8, no solubility in water, a high resistance to chemical attack, a decomposition temperature of 500° to 700° C and a Knoop hardness of 100-120.

NACRITE - is the mineral form of $Al_2O_3 \cdot 2SiO_2 \cdot 2H_2O$. It is a member of the kaoli-nite group of clays and has a thin hexagonal plate like structure.

NANTOKITE - is the mineral form of CuCl. See the data section for the properties.

NEPHELINE - is the mineral form of $K_2O \cdot 3Na_2O \cdot 4Al_2O_3 \cdot 9SiO_2$. It has a hexago-nal structure, a specific gravity of 2.5 and a hardness of 5.5-6 Mohs. It converts to carnegiete at 1254° C.

NEPHELINE SYENITE - consists mainly of nepheline, microcline and albite. It is used as a replacement for feldspar.

NEPHELITE - is the mineral form of $N_2O \cdot Al_2O_3 \cdot 2SiO_2$. It has a hexagonal struc-ture, an index of refraction of 1.5, a specific gravity of 2.6, a melting temperature of 1526° C and no solubility in water.

NESQUEHONITE - is the mineral form of $MgCO_3 \cdot 3H_2O$. It has a rhombic struc-ture, an index of refraction of 1.5, a specific gravity of 1.9, a melting temperature of 165° C and a solubility in water of 0.2 g/100ml.

NETWORK-FORMING - are glasses that display a structure which is chemically bonded throughout but not crystalline.

NETWORK-MODIFYING - are atoms which alter the network structure when they are added to a glass.

NEUTRAL VERDIGRIS - is the mineral form of $Cu(C_2H_3O_2)_2 \cdot H_2O$. It has an index of refraction of 1.6, a specific gravity of 1.9, a melting temperature of 115° C, a decomposition temperature of 240° C, a solubility in cold water of 7.2 g/100ml and in hot water of 20 g/100ml. Neutral verdigris is used as a green pigment.

NEWBERYITE - is the mineral form of $MgHPO_4{\cdot}3H_2O$. It has a rhombic structure, an index of refraction of 1.5, a specific gravity of 2.1 and a slight solubility in water.

NEWTONITE - is the mineral form of $Al_2O_3{\cdot}2SiO_2{\cdot}5H_2O$.

NICCOLITE - is the mineral form of NiAs. It has a hexagonal structure, a specific gravity of 7.6, a melting temperature of 968° C and no solubility in water.

NITROBARITE - is the mineral form of barium nitrate. It has a cubic structure, an index of refraction of 1.6, a specific gravity of 3.2, a melting temperature of 592° C, a solubility in cold water of 9 g/100ml and in hot water of 34 g/100ml.

NON-BRIDGING -is an atom that does not continue the network structure in a glass.

NONTRONITE - is the mineral form of $Fe_2O_3{\cdot}3SiO_2{\cdot}xH_2O$.

OCTAHEDRITE - is an alternate term for anatase.

OLDHAMITE - is the mineral form of calcium sulfide. It has a cubic structure, an index of refraction of 2.1, a specific gravity of 2.5, a solubility in cold water of 0.021 g/100ml. and in hot water of 0.048 g/100ml.

OLIGOCLASE - is a mixture of albite and anorthite.

OLIVINE - is a green mineral containing a mixture of fersterite and tayalite. The softening point is 1700° C.

OPACITIERS - are additives to a glaze to render it opaque. This is accomplished by having the additive as an irregularly shaped particulate with an index of refraction different from the glaze.

OPAL - is the mineral form of $SiO_2{\cdot}xH_2O$. It has an amorphous structure, an index of refraction of 1.4-1.5, a specific gravity of 2.2, a melting temperature under 1600° C and no solubility in water.

ORPIMENT - is the mineral form of arsenic sulfide. See the data section for the properties.

ORTHOCLASE - is a mineral form of potassium aluminosilicate. See the data section for the properties.

ORTHOENSTATITE - is the mineral form of $MgO \cdot SiO_2$. It converts to clinoenstatite at 1140° C.

PANDERMITE - is a mineral of the approximate form $4CaO \cdot 5B_2O_3 \cdot 7H_2O$. When it is used in a glaze it does not require fritting because it is insoluble in water.

PARALARIONITE - is the mineral form of $PbCl_2 \cdot PBO \cdot H_2O$. It has a monoclinic structure, an index of refraction of 2.1, a specific gravity of 6.1 and a decomposition temperature of 150° C.

PARIS GREEN - has a composition of $Cu(C_2H_3O_2)_2 \cdot 3Cu(AsO_2)_2$. It is used as a green pigment which is insoluble in water.

PBN - is an acronym for pyrolytic boron nitride.

PERICLASE - is a mineral form of magnesium oxide. See the data section for the properties.

PERLITE - is a siliceous volcanic glass used for thermal and sound insulation.

PEROVSKITE - are oxides with the stoichiometry of ABO_3. The A atoms are physically much larger than the B atoms which results in a cubic structure. Perovskites are generally used as the crystalline component in a glass-ceramic to promote electrical conductivity.

PEROWSKITE - is the mineral form of $CaO \cdot TiO_2$. It has a cubic or rhombic structure, an index of refraction of 2.3, a specific gravity of 4.1, a Mohs hardness of 5.5 and a melting temperature of 1975° C.

PEST BEHAVIOR - is catastrophic oxidation that occurs within the mid-range of the solid state. It is usually a region of a few hundred degrees which results from the ceramic being warm enough to be susceptible to oxidation but not yet able to form the protective coating it will at a higher temperature.

PETALITE - $(Li,Na)_2O \cdot Al_2O_3 \cdot 8SiO_2$, has a monoclinic structure, a specific gravity of 2.4 and a hardness of 6.5 Mohs.

PHENAZITE - is the mineral form of Be_2SiO_4. It has a triclinic structure, an index of refraction of 1.7 and a specific gravity of 3.0

PHOLERITE - $Al_2O_3 \cdot 3SiO_2 \cdot 4H_2O$

PHOSPHATE PENTOXIDE - P_2O_5. It is primarily used as a glass additive to aid in light transmission. It crystallizes into hexagonal, orthorhomic, and tetragonal as well as forming a glass.

PINNOITE - is the mineral form of $Mg(BO_2)_2 \cdot 3H_2O$. It has a tetragonal structure, an index of refraction of 1.6 and a specific gravity of 2.3.

PLAGIOCLASE - $NaKO \cdot Al_2O_3 \cdot 6SiO_2$

PAGIOCLASE FELDSPAR - is composed of albite and anorthite.

PLASTER OF PARIS - is the common term for hemihydrate.

PLASTICITY - has the usual meaning in mechanical properties. However, it is also used extensively in the ceramics industry relative to clays. Clay generally becomes plastic when water is added. They can then slip plastically into a new shape which will be retained when the shaping pressure is removed. This new shape will continue be retained when the water is removed although shrinking may occur.

POLLUCITE - has the chemical composition $2Cs_2O \cdot 2Al_2O_3 \cdot 9SiO_2 \cdot H_2O$ or $Cs_2O \cdot 2Al_2O_3 \cdot 4SiO_2$ or $(Cs\backslash Na)_2O \cdot Al_2O_3 \cdot 5SiO_2 \cdot H_2O$ It has a cubic crystalline structure and is colorless.

PORCELAIN - is composed of crystalline quartz and mullite dispersed in a matrix of glass with a porosity of 5% or less. Clay, flint and feldspar are usually mixed together and fired at 1100° to 1400° C to make porcelain.

PORTLAND CEMENT - is the most widely used form of cement. It is made from a chalk or limestone plus a clay or shale. The minerals are sintered and ground to form clinker. An addition of 3%-5% calcium sulfate is then made.

POTASH SPAR - is primarily orthoclase in composition.

POTASSIUM ALUMINOSILICATE - $K_2O \cdot Al_2O_3 \cdot 6SiO_2$. See the data section for the properties.

POTASSIUM CARBONATE - K_2CO_3. It has a specific gravity of 2.4, a melting temperature of 891° C and a solubility in water of 112 g/100ml at 20° C and 156 g/100ml at 100° C. The main use is as a flux in glazes.

POTASSIUM FELDSPAR - is a feldspar rich in potassium aluminosilicates.

POTASSIUM HYDROXIDE - attacks all ceramics to some extent, especially in the molten form. Zirconia and spinel are both reasonably resistant.

POTASSIUM NITRITE - KNO_3. It has a specific gravity of 2.1, a phase change at 129° C, a melting temperature at 334° C and a decomposition temperature of 400° C. Its solubility in water is 13.3 g/100ml at 0° C, 31.6 g/100ml at 20° C and 247 g/100ml at 100° C. It is a strong oxidizing agent up to decomposition.

POTASSIUM OXIDE - K_2O. It has a cubic structure, a specific gravity of 2.3, a decomposition temperature of 350° C and a high solubility in water.

POWELLITE - is the mineral form of calcium molybdate. It has a tetragonal structure, an index of refraction of 2.0, a specific gravity of 4.4-4.5, and no solubility in water.

POZZOLANAS - are a broad variety of materials which react with lime and water to form cement. Sources include volcanic earth, diatomaceous earth, oil shales and fuel ash. They conform to ASTM standard C6 18-80.

POZZOLANIC CEMENT - is composed of up to 40% pozzolana with the remainder being conventional cement. It is more resistant to sea water, sulfates and acidic water than Portland cement.

PROUSTITE - is the mineral form of Ag_3AsS_3. It has an index of refraction of 2.8-3.0, a trigonal structure, a specific gravity of 5.5, a melting temperature of 490° C and no solubility in water.

PSEUDOCOTUNNITE - is the mineral form of $2KCl \cdot PbCl_2$. It has a melting temperature of 490° C and is soluble in water.

PSEUDOWOLLASTONITE - is the mineral form of calcium metasilicate. It has a monoclinic structure, an index of refraction of 1.6-1.7, a specific gravity of 2.9, a melting temperature of 1540° C, a solubility in water of 0.01 g/100ml. Pseudowollastonite converts from wollastonite at 1125° C.

PSZ - is an acronym for partially stabilized zirconia.

PUCHERITE - is the mineral form of $Bi_2O_3 \cdot V_2O_5$. It has a rhombic structure, a specific gravity of 6.3 and an index of refraction of 2.4-2.5.

PUMICE - is a replacement for feldspar when the final product is not white or clear.

PYRARGYRITE - is the mineral form of Ag_3SbS_3. It has an index of refraction of 3.1, a trigonal structure, a specific gravity of 5.8, a melting temperature of 486° C and no solubility in water.

PYREX - is a glass with a high silicon dioxide content, low thermal expansion, high softening point and good chemical resistance. A common composition is 80.5% silica, 11.8% boron oxide, 2.0% alumina, 0.7% arsenic trioxide, 0.3% calcium oxide, 0.1% magnesium oxide, 0.2% potassium oxide, and 4.4% sodium oxide.

PYRITE - is the mineral form of iron disulfide. See the data section for the properties.

PYROCHROITE - is the mineral form of $Mn(OH)_2$. It has an index of refraction of 1.7, a trigonal structure, a specific gravity of 3.3 and a solubility in water of 0.0002 g/100ml.

PYROLITIC GRAPHITE - is graphite formed by chemical vapor deposition. The resulting material is very pure, has low inherent stresses and has a cone macrostructure.

PYROMETRIC CONE EQUIVALENT - is the temperature at which a ceramic attains a specific viscosity. It is often used in place of softening or melting temperature.

PYROLLUSITE - is the mineral form of manganese dioxide. It has a rhombic structure, a specific gravity of 5.0 and no solubility in water.

PYROPHANITE - is the mineral form of $MnTiO_3$. It has an index of refraction of 2.5, a trigonal structure, a specific gravity of 4.5 and a melting temperature of 1360° C.

PYROPHYLLITE - is the mineral form of $Al_2O_3 \cdot 4SiO_2 \cdot H_2O$ It has an abrasive character which is resistant to thermal shock due to a good thermal conductivity and a low coefficient of thermal expansion.

PZT - is the acronym for lead zirconate titanate. This material is used extensively in piezoelectric transducers despite its dielectric constant of 1800 and its specific gravity of 7.9

QUARTZ - is a mineral form of silicon dioxide. See the data section for the properties.

RASORITE - is an alternate name for kernite.

RASPITE - is a mineral form of $PbWO_4$. It has a monoclinic structure, an index of

249

refraction of 2.3, a melting temperature of 1123° C and a solubility in water of 0.04 g/100ml. It has the same chemical composition as stolzite.

RBSN - is an acronym for reaction bonded silicon.

REACTION-BONDING - is a technique used primarily for silicon based ceramics. At high temperature, silicon is reacted with an appropriate non-metal which results in a chemically bonded ceramic.

REALGAR - is the mineral form of arsenic disulfide. It has a monoclinic structure, an index of refraction of 2.5-2.6, a specific gravity of 3.5 for the alpha form and 3.3 for the beta form, the alpha to beta transformation temperature is 267° C, the melting temperature is 307° C and the boiling temperature is 565° C. Light is transmitted through a 2mm thickness of realgar in excess of 10% in the region of 0.6-3 microns.

RECRYSTALLIZATION - is a high temperature rearrangement of polycrystalline material. Any deformed crystals, usually produced by mechanical working, are at a high energy level and will rearrange themselves into smaller more prefect crystals if conditions are favorable. Normally the added energy needed to allow for the diffusion necessary for rearrangement is heat.

REDDINGITE - is the mineral form of $Mn_3(PO_4)_2 \cdot 3H_2O$. It has a rhombic structure, an index of refraction of 1.7, and a specific gravity of 3.1.

RED LEAD - is the mineral form of Pb_3O_4. It has a specific gravity of 9.1, decomposes to lead oxide at 500° C., and use primarily as a flux in glazes.

REFRACTORY - is a ceramic material that is resistant to melting until high temperatures. This does not imply strength other than shape retention.

REFINE - is the process of removing the gas bubbles in a glass.

REINFORCED CONCRETE - is concrete with added material to increase the overall strength. Usually steel rods are used as the reinforcing material.

RHODONITE - is the mineral form of $MnSiO_3$, It has a triclinic structure, an index of refraction of 1.7, a specific gravity of 3.7, a melting temperature of 1323° C and no solubility in water.

RHOMBOKLAS - is the mineral form of $Fe_2O_3 \cdot 4SO_3 \cdot 9H_2O$. It has an index of refraction of 1.5-1.6, a specific gravity of 2.2 and high solubility in water.

RICKARDITE - is the mineral form of Cu_4Te_3. It has a tetragonal structure and a specific gravity of 7.5.

RODOCHROSITE - is the mineral from of $MnCO_3$. It has a rhombic structure, a specific gravity of 3.1 and a solubility in water of 0.007 g/l00mi.

ROESSLERITE - is the mineral form of $MgHAsO_4 \cdot 7H_2O$. It has a monoclinic structure and a specific gravity of 1.9

RUBY - is a mineral form of aluminum oxide. See the data section for the properties.

RUTILE - is the mineral form of titanium dioxide. See the data section for the properties.

SALT CAKE - is the common term in the glass industry for sodium oxide.

SAPONITE - has a chemical structure of $2MgO \cdot 3SiO_2 \cdot xH_2O$. It is a member of the montmorillonite group of clays.

SAPPHIRE - is a form of corundum. Large clear pieces are deep blue and used as a gem. See the data section under aluminum oxide for the properties.

SAUCONITE - has a chemical composition of $2ZnO \cdot 3SiO_2 \cdot xH_2O$. It is a member of the montmorillonite group of clays.

SCAMING - is the reaction product formed when glass is exposed to water. This is a surface phenomenon resulting in a haze on the glass.

SCHEELITE - is the mineral form of $CaWO_4$. It has a tetragonal structure, an index of refraction of 1.9, a specific gravity of 6.1 and a solubility in water of 0.2 g/100ml. It is used in the production of tungsten carbide.

SCHOTTKY DEFECT - is a combination of vacancies that constitute a stoiciometric group. These atoms move out of the ceramic independently to electrically unbalanced areas like grain boundaries.

SCHROETTERITE - $8Al_2O_3 \cdot 3SiO_2 \cdot 3H_2O$

SCHULTENITE - is the mineral form of $PbHAsO_4$. It has a monoclinic structure, a specific gravity of 5.8, a decomposition temperature of 720° C and no solubility in water.

SECOND ORDER TRANSFORMATION - is a structural change that takes place within a material without a discontinuous change in volume.

SEEDS - are small bubbles which develop during glass formation. They are usually removed from the molten glass by a fining agent which gives off large bubbles of oxygen. As the oxygen bubbles rise in the melt they sweep out the seeds.

SELLAITE - is the mineral form of MgF_2. It has a tetragonal structure, an index of refraction of 1.4, a specific gravity of 3.1, a melting temperature of 1261° C, a boiling temperature of 2239° C and a solubility in water of 0.008 g/100ml.

SENARMONTITE - is the mineral form of antimony trioxide. See the data section for the properties.

SHORT GLASS - has large variation of viscosity as a function of temperature which results in a short workable region.

SIALON - or Si-Al-O-N, refers to the ceramics based on silicon, aluminum, oxygen and nitrogen compounds.

SIDERITE - is the mineral form of iron carbonate. It has an index of refraction of 1.6-1.9, a trigonal structure, a specific gravity of 3.8 and a solubility in water of 0.007 g/100ml.

SIDEROTIL - is the mineral form of $FeSO_4 \cdot 5H_2O$. It has a triclinic structure, an index of refraction of 1.5, a specific gravity of 2.2 and a high solubility in water.

SIERRALITE - is a mineral composed mainly of silica, alumina and magnesia.

SILICA - is an alternate name for silicon dioxide.

SILICON CARBIDE - SiC. See the data section for the properties.

SILICON DIOXIDE - SiO_2. See the data section for the properties.

SILICON NITRIDE - Si_3N_4. See the data section for the properties.

SILLIMANITE - is the mineral form of aluminum silicate. See the data section for the properties.

SILVER BROMIDE - The mineral form is bromyrite.

SILVER CHLORIDE - The mineral form is cerargyrite.

SILVER IODIDE - The mineral form is iodyrite.

SILVER NITRATE - $AgNO_3$. See the data section for the properties.

SINTERING - is the process of converting individual particles of ceramic into a cohesive material. The loose particles are pressed together in a mold. The particles have a high surface area which serves as an energy source to drive rearrangement via diffusion. Over time and especially with the addition of heat the surfaces in intimate contact broaden and become chemically bonded.

SLAG CEMENT - is made from the slag of a blast furnace. This contains the necessary chemistry to produce cement but previous attempts to use it have not always proven successful. The primary problem seems to be the crystallization of the slag. Even when crystallization has been avoided the slag does not make a suitable cement until the surface has been chemically activated. In practice the slags are quenched, finely ground and then mixed with Portland cement. The Portland cement both contributes to the strength and activates the slag.

SLIP - is a mixture of clay and water, usually in a very fluid form.

SLUMP - is the flow characteristics of cement slurry.

SLURRY - is an alternate term for slip.

SMITHSONITE - is the mineral form of $ZnCO_3$. It has an index of refraction of 1.6-1.8, a trigonal structure, a specific gravity of 4.4, and a solubility in water of 0.001 g/100ml.

SODA ASH - is an alternate name for sodium carbonate.

SODA SPAR - is an alternate term for albite.

SODIUM CARBONATE - Na_2CO_3. It has a specific gravity of 2.5, a melting temperature of 851° C, a solubility in water of 7.1 g/100ml at 0° C and 45.5 g/100ml at 100° C. It is used as a deflocculant in clay.

SODIUM CHLORIDE - NaCl. The mineral form is halite.

SODIUM FELDSPAR - is an alternate term for albite.

SODIUM HYDROXIDE - NAOH. It has a specific gravity of 2.1, a melting temperature of 318° C, a boiling temperature of 1390° C, a solubility in cold water of 42 g/100ml and in hot water of 347 g/100ml. It attacks all ceramics to some extent, especially in the molten state, but alumina is moderately resistant to it. It is commonly used as a deflocculant in clay.

SODIUM NITRITE - $NaNO_2$. It is used in glass making both as a source of soda and as an oxidizing agent. It has a rhombic structure, a specific gravity of 2.2, a melting temperature of 271° C, a decomposition temperature of 320° C, a solubility in cold water of 82 g/100ml and in hot water of 163 g/100ml.

SODIUM OXIDE - Na_2O. It has a specific gravity of 2.3,a sublimation temperature of 1275° C and a dilute solubility in water.

SOFTENING POINT - is the temperature at which a glass fails to support its own weight without deformation.

SPALLING - is cracking on cooling due either to thermal shock or a difference in thermal contraction between the ceramic and its glaze.

SPERRYLITH - is the mineral form of $PtAs_2$. It has a cubic structure, and a specific gravity of 11.8

SPHALERITE - is the mineral form of beta ZnS. It has a cubic structure, an index of refraction of 2.4, a specific gravity of 4.1, a transformation to wurtzite at 1020° C and a slight solubility in water.

SPHEROCOBALTITE - is the mineral form of cobalt carbonate. It has an index of refraction of 1.9, a trigonal structure, a specific gravity of 4.1, and no solubility in water.

SPINEL - is a general cubic crystalline type which has the formula AB_2O_4 where A and B are metals. Within the spinel structure are the normal and the inverted structures differing by the way in which the metals occupy sites. It is also used to specifically mean magnesium aluminate.

SPODUMENE - has the chemical composition of $Li_2O \cdot Al_2O_3 \cdot 4SiO_2$. See the data section for the properties.

STEATITE - has the chemical composition $4MgO \cdot 5SiO_2 \cdot H_2O$. See the data section for the properties.

STEVENSITE - seems to be an interlayered clay of talc and saponite.

STIBNITE - is the mineral form of antimony trisulfate. It has a rhombic structure, an index of refraction of 3.2-4.3, a specific gravity of 4.6, a melting temperature of 550° C, a boiling temperature of 1150° C and a solubility in water of 0.0002 g/100ml.

STOICHIOMETRY - is the formation of ceramic materials from atoms that come together in small whole number ratios. An example of this is silicon dioxide which has one silicon to every two oxygen atoms throughout its entire structure.

STOLZITE - is the mineral form of $PbWO_4$. It has a tetragonal structure, an index of refraction of 2.2-2.3, a specific gravity of 8.2, no solubility in water, and the same chemical composition as raspite.

STONEWARE CLAYS - resemble ball clays except they do not burn to be white. The burning temperature is about 1100° C.

STRAIN TEMPERATURE - is the maximum service temperature of a glass. Just how much elongation the glass undergoes at this temperature is measured by ASTM C336.

STRONTIANITE - is the mineral form of strontium carbonate. It has a rhombic structure which transforms to hexagonal at 926° C, an index of refraction of 1.5-1.7, a specific gravity of 3.7, a melting temperature of 1497° C, a solubility in cold water of 0.001 g/100ml and in hot water of 0.7 g/100ml.

STRONTIUM CARBONATE - $SrCO_3$. It decomposes to SrO at 1340° C.; it is used as a glaze.

STRONTIUM SULFATE - $SrSO_4$. The mineral form is celestite.

SULFATE RESISTANT PORTLAND CEMENT - has the tricalcium aluminate reduced from 8%-11% to 3%-4 %, and it conforms to ASTM C150-81 standard.

SULFUR CONCRETE - is a composite of aggregate held together by sulfur. Usually the sulfur matrix has a plasticizer as an additive.

SULFUR TRIOXIDE - SO_3. It has a specific gravity of 2.0, a melting temperature of 17° C, and a boiling temperature of 45° C. It is used as an additive in crown glass.

SUPERSULFATED CEMENTS - has the composition 80%-85% granulated blast furnace slag, 10%-15% anhydrous calcium sulfate and 5% Portland cement or hydrated lime. The set cement is resistant to sulfate solutions and acids.

SYCOPORITE - is the mineral form of CoS. It has a specific gravity of 5.5, a melting temperature of 1116° C, and a solubility in water of 0.0004 g/100ml.

SYLVITE - is the mineral form of KCl, it has a cubic structure, an index of refraction of 1.5, a specific gravity of 2.0, melting temperature of 770° C, a sublination temperature of 1500° C, a solubility in cold water of 24 g/100ml and in hot water of 57 g/100ml.

SZMIKITE - is the mineral form of $MnSO_4 \cdot H_2O$. It has a monoclinic structure, an index of refraction of 1.6, a specific gravity of 3.0, a solubility in cold water of 99 g/100ml. and in hot water of 80 g/100ml.

255

SZOMOLNIKITE - is the mineral form of $FeSO_4·H_2O$. It has a monoclinic structure, a specific gravity of 3.0 and is slightly soluble in water.

TALC - is the mineral form of $3MgO·4SiO_2·H_2O$. It possesses a layer structure similar to clay and is used in tile manufacture. The melting temperature is 1490° C, the specific gravity is 2.6-2.8 and the hardness is 1-2 Mohs.

TANTALUM BROIDE - TaB_2. See the data section for the properties.

TANTALUM CARBIDE - TaC. See the data section for the properties.

TAPIOLITE - is the mineral form of $Fe(TaO_3)_2$. It has a tetragonal structure, an index of refraction of 2.3 and a specific gravity of 7.3

TARAPACAITE - is the mineral form of K_2CrO_4. It has a rhombic structure, a specific gravity of 2.7, a melting temperature of 968° C, a solubility in cold water of 639 g/100ml and in hot water of 790 g/100ml.

TENORITE - is the mineral form of CuO. It has a monoclinic structure, a specific gravity of 6.3-6.5, a melting temperature of 1326° C and no solubility in water.

TERRACOTTA - is an unglazed porous ceramic often made of red clay.

TETRADYMITE - is the mineral form of Bi_2Te_3. It has a specific gravity of 7.7 and a melting temperature of 573° C.

THALLIUM CHLORIDE - TlCl. See the data section for the properties.

THENARD'S BLUE - $CoAl_2O_4$. It has a cubic structure, is insoluble in water and is used as a blue pigment.

THENARDITE - is the mineral form of Na_2SO_4. It has an orthorhombic structure, an index of refraction of 1.5, a specific gravity of 2.7, a solubility in cold water of 5 g/100ml. and in hot water of 43 g/100ml.

THIXOTROPY - is the property of some clays wherein their viscosity is dependent on prior flow history.

THORIA - is an alternate name for thorium dioxide.

THORINAITE - is the mineral form of thorium dioxide. See the data section for the properties.

THORIUM CARBIDE - ThC_2. It has a thermal conductivity of 0.08 W/cmK at 200° to 400° C, a tetragonal structure, a specific gravity of 9, a melting temperature of 2655° C, a boiling temperature of about 5000° C and a dilute solubility in water.

THORIUM DIOXIDE - ThO_2. See the data section for the properties.

TIEMANNITE - is the mineral form of HgSe. It has a specific gravity of 8.3 and no solubility in water.

TINCAL - is an alternate name for borax.

TIN OXIDE - SnO_2. See the data section for the properties.

TITANIA - is an alternate name for titanium dioxide.

TITANITE - is the mineral form of $TiO_2 \cdot CaO \cdot SiO_2$. It has a specific gravity of 3.5 and a Mohs hardness of 5-5.5

TITANIUM CARBIDE - TiC. See the data section for the properties.

TITANIUM CARBONITRIDE - TiCN. It has properties similar to titanium nitride.

TITANIUM DIBORIDE - TiB_2. See the data section for the properties.

TITANIUM DIOXIDE - TiO_2. See the data section for the properties.

TITANIUM NITRIDE - TiN. See the data section for the properties.

TITANIUM OXIDE - is an alternate name for titanium dioxide.

TOPAZ - is the mineral form of aluminum fluosilicate. It has a rhombic structure, a of hardness 8 Mohs, an index of refraction of 1.6 and a specific gravity of 3.4-3.6 Topaz converts to mullite and quartz on heating above 1090° C.

TRASS CEMENT - is an alternate term for pozzolana cement.

TREMOLITE - is a form of asbestos. See the data section under asbestos for the properties.

TRICALCIUM PHOSPHATE - $Ca_3(PO_4)_2$. It has a specific gravity of 3.14, a melting temperature of 1670° C and a solubility in water of 0.023 g/100ml. It decomposes in hot water.

TRIDYMITE - is a mineral form of silicon dioxide. See the data section for the properties.

TROILITE - is the mineral form of FeS. It has a hexagonal structure, a specific gravity of 4.7, a melting temperature of 1193° to 1199° C and a solubility in water of 0.0006 g/100ml.

TUNGSTEN CARBIDE - WC. See data section for the properties.

TUNGSTEN DISULFIDE - WS_2. See data section for the properties.

TUNGSTENITE - is the mineral form of tungsten disulfide. See the data section for the properties.

URANIUM CARBIDE - UC, UC_2, or U_2C_3. See the data section for properties.

URANIUM NITRIDE - UN. See the data section for properties.

URANIUM OXIDE - UO_2. See the data section for properties.

URANIUM SULFATE - US. See the data section for properties.

VALENTINITE - is the mineral form of antimony trioxide. See data section for the properties.

VANADIUM CARBIDE - VC. See the data section for the properties.

VERMICULITE - is the mineral form of $(OH)_2(MgFe)_3(SiAlFe)_4O_{10}\cdot 4H_2O$. It has a melting temperature of 1350° C, a stable working temperature of 1100° C and is used in acoustical tiles as well as thermal insulators.

VILLIAUNITE - is the mineral form of NaF. It has a cubic or tetragonal structure, an index of refraction of 1.3, a specific gravity of 2.6, a melting temperature of 993° C, a boiling temperature of 1695° C and a solubility in water of 4.2 g/100ml.

VITREOUS CARBON - is manufactured by degradation of cross-linked polymeric materials into pure carbon in the glassy state.

VITREOUS SILICA - is silicon dioxide in the glassy state. See data section for the properties.

VITRIFICATION - is the processing of a ceramic into a glassy state.

VIVIANITE - is the mineral form of $Fe_3(PO_4)_2\cdot 8H_2O$. It has a monoclinic structure, an index of refraction of 1.6, a specific gravity of 2.6 and no solubility in water.

258

VOLCANIC ASH - is an alternate term for pumice.

WATER GLASS - is a glass with a composition $NaO \cdot xSiO$ where x varies from 3 to 5. It is slightly solubile in water, and it used as a deflocculant in clay.

WAVELLITE - has the chemical composition $Al_6 \cdot (OH)_6 \cdot (PO_4)_4 \cdot 9H_2O$

WHISKERS - are single crystal filament ceramics useful for their purity, large surface area and known crystallographic orientation.

WHITE MICA - is the common term for muscovite.

WHITLOCKITE - is the mineral form of $Ca_3(PO_4)_2$. It has an index of refraction of 1.6, a specific gravity of 3.1, a melting temperature of 1670° C, and a solubility in cold water of 0.002 g/100ml.

WILLEMITE - is the mineral form of Zn_2SiO_4. It has an index of refraction of 1.7, a trigonal structure, a specific gravity of 4.1, a melting temperature of 1509° C and no solubility in water.

WINDOW GLASS - has a common composition of 72.1% silica, 1.1% alumina, 0.2% ferric oxide, 10.2% calcium oxide, 6.2% magnesia and 13.6% sodium oxide.

WITERITE - is the mineral form of $BaCO_3$. It has a rhombic structure and an index of refraction of 1.7.

WOLFRAMITE - is the mineral $(Fe,Mn)WO_4$. It is used for the production of tungsten carbide.

WOLLASTONITE - is the mineral form of calcium meta silicate. It has a fibrous structure, an index of refraction of 1.6, a specific gravity of 2.9, a melting temperature of 1540° C, a hardness of 4.5-5 Mohs., a coefficient of thermal expansion of $9.4X10^{-6}$/K over 100° to 200° C, and a solubility in water of 0.01 g/100ml. It converts to pseudowollastonite at 1125° C.

WUESTITE - is the mineral form of FeO. It has a cubic structure, an index of refraction of 2.3, a specific gravity of 5.7, a melting temperature of 1369° C and no solubility in water. Wuestite is often an undersirable coloring agent in glass.

WULFENITE - is the mineral form of $PbMoO_4$. It has a tetragonal structure, a specific gravity of 6.9, a melting temperature of 1060° C and no solubility in water.

WURTZITE - is the mineral form of alpha zinc sulfide. It has a hexagonal structure, an index of refraction of 2.4, a specific gravity of 4.0, a melting temperature of 1700° C and a solubility in water of 0.0007 g/100ml. The amount of light transmitted through a 2 mm. thickness is in excess of 10% for the region 0.6-14.5 microns. It is used as a pigment.

YAG - is an acronym for $5Al_2O_3 \cdot 3Y_2O_3$

YAM - is an acronym for $Al_2O_3 \cdot 2Y_2O_3$

YTTRIA - is an alternate term for yttrium oxide.

YTTRIUM OXIDE - Y_2O_3. See data section for properties.

ZINCITE - is the mineral form of zinc oxide. See the data section for the properties.

ZINKOSITE - is the mineral form of $ZnSO_4$. It has a rhombic structure, an index of refraction of 1.7, a specific gravity of 3.5, a decomposition temperature of 600° C and a high solubility in water.

ZINNWALDITE   -   is   the   mineral   form   of $2Li_2O \cdot K_2O \cdot 2(Fe,Mg)\dot{O} \cdot 2Al_2O_3 \cdot 6SiO_2 \cdot F_2 \cdot H_2O$. It has a monoclinic structure and a specific gravity of 2.9-3.1

ZINC OXIDE - ZnO. See the data section for the properties.

ZINC SULFIDE - ZnS. The mineral form is wurtzite.

ZIRCON - $ZrO_2 \cdot SiO_2$. See the data section for the properties.

ZIRCONIA - is an alternate term for zirconium oxide.

ZIRCONIUM DIBORIDE - ZrB_2. See data section for the properties.

ZIRCONIUM CARBIDE - ZrC. See the data section for the properties.

ZIRCONIUM NITRIDE - ZrN. It has a specific gravity of 7.1, a melting temperature of 2980° C, no solubility in water and is extremely stable.

ZIRCONIUM OXIDE - ZrO_2. See the data section for the properties.

ZIRKITE - is a mineral containing about 75% zirconium oxide.

ZTA - is an acronym for zirconia toughened with alumina.

BIBLIOGRAPHY

"Metals and Ceramics Information Center Handbook." # MCIC-HB-07 volume I. Columbus: Battelle, 1976.

"Metals and Ceramics Information Center Handbook." # MCIC-HB-07 volume II, Columbus: Battelle, 1979.

"Metals and Ceramics Information Center Handbook." # MCIC-HB-07 volume III, Columbus: Battelle, 1981.

Aben and Guillemet. *Photoelasticity of Glass.* New York: Springer-Verlag, 1993.

Acquaviva and Bortz, eds. *Structural Ceramics and Design.* New York: Gordon and Breach, 1969.

American Ceramic Society. *Ceramic Materials and Components for Engines.* Westerville: American Ceramic Society, 1989.

Balta, P. and E. Balta. *Introduction to the Physical Chemistry of the Vitreous State.* Kent: Abacus Press 1976.

Baumgart, Dunham, and Amstutz. *Process Mineralogy of Ceramic Materials.* New York: Elsevier, 1984.

Blachere and Pettit. *High Temperature Corrosion of Ceramics.* Park Ridge: Noyes, 1989.

Bradt and Tressler, eds. *Deformation of Ceramic Materials.* New York: Plenum Press, 1975.

Briant, Petrovic, Bewlay, Vasudevan, and Lipsitt. *High Temperature Silicates and Refractory Alloys.* Pittsburgh: Materials Research Society, 1994.

Broutman and Krock, eds. *Modern Composite Materials.* Reading: Addison-Wesley, 1967.

Budworth, D. W. *An Introduction to Ceramic Science.* Oxford: Pergamon Press, 1970.

Bunn, C. W. *Chemical Crystallography.* Oxford: Oxford University Press, 1961.

Burke, Gorum, and Katz, eds. *Ceramics for High-Performance Applications.* Chestnut Hill: Brook Hill Publishing Company, 1974.

Burke, J. E., ed. *Progress in Ceramic Science.* Volume 4. New York: Pergamon Press, 1966.

Burke, Lenoe, and Katz, eds. *Ceramics for High Performance Applications II.* Chestnut HIll: Brook Hill Publishing Company, 1978.

Clancy, T. A. "High-Temperature Corrosion Resistance of Ceramic Materials." Information Circular 8843. Washington: United States Department of the Interior Bureau of Mines, 1981.

Davidge, R. W. *Mechanical Behaviour of Ceramics.* Cambridge: Cambridge University Press, 1979.

Davidge, R. W., ed. *Engineering With Ceramics.* Stoke-on-Trent: British Ceramic

Society, 1982.

Doremus, R. H. *Glass Science.* New York: J Wiley and Son, 1994.

Dorre and Hubner. *Alumina Processing: Properties, and Applications.* New York: Spring-Verlag, 1984.

Dotsenko, Glebov, and Tsekhomsky. *Physics and Chemistry of Photochromic Glasses.* New York: CRC Press, 1998.

Eglington, M. S. *Concrete and its Chemical Behaviour.* London: Thomas Telford, 1987.

Fletcher, N. H. *The Chemical Physics of Ice.* London: Cambridge University, 1970.

Frankhouser, Brendley, Kieszek, and Sullivan. *Gasless Combustion Synthesis of Refractory Compounds.* Park Ridge: Noyes, 1985.

Frankhouser, W. L. *Advanced Processing of Ceramic Compounds.* Park Ridge: Noyes, 1987.

Frechette, Pye, and Rase., eds. *Quality Assurance in Ceramic Industries.* New York: Plenum Press, 1978.

Fulrath and Pask, eds. *Ceramic Microstructures.* New York: John Wiley and Sons, 1968.

Gitzen, Walter, ed. *Alumina as a Ceramic Material.* Columbus: The American Ceramic Society, 1970.

Glasser and Potter, eds. *High Temperature Chemistry of Inorganic and Ceramic Materials.* London: Burlington House, 1977.

Godfrey, D. J., ed. *The Mechanical Engineering Properties and Applications of Ceramics.* Stoke-on-Trent : British Ceramic Society, 1978.

Gorsler, F. W., ed. *Cutting Tool Materials.* Metals Park: ASM, 1980.

Greenwood, N. N. *Ionic Crystals Lattice Defects and Nonstoichiometry.* London: Butterworths, 1968.

Hamilton, David. *Manual of Architectural Ceramics.* London: Thames and Hudson, 1978.

Holden, Robert. *Ceramic Fuel Elements.* New York: Gordon and Breach, 1966.

Holliday, Leslie, ed. *Composite Materials.* New York: Elsevier Publishing Company, 1966.

Hove and Riley, eds. *Ceramics for Advanced Technologies.* New York: John Wiley and Sons, 1965.

Jona and Shirane. *Ferroelectric Crystals.* New York: Dover Publications Inc, 1993.

King, Alan, and W. M Wheildon. *Ceramics in Machining Processes.* New York: Academic Press, 1966.

Kingery, Bowen, and Uhlmann. *Introduction to Ceramics.* New York: John Wiley and Sons, 1976.

Kirchner, Henry. *Strengthening of Ceramics.* New York: Marcel Dekker, 1979.

Klingsberg, Cyrus, ed. *The Physics and Chemistry of Ceramics.* New York: Gordon and Breach, 1963.

Kosolapova, T. Y., ed. *High Temperature Compounds: Properties, Production, Applications.* New York: Hemisphere Publishing Corp, 1990.

Krockel, Mertz, and Van Der Biest, eds. *Ceramics in Advanced Energy Technologies.* Boston: D. Reidel Publishing Company, 1984.

Lay, Lewis. *Corrosion Resistance of Mechanical Ceramics.* London: Her Majesty's Stationery Office, 1983.

Lenoe, Katz, and Burke, eds. *Ceramics for High-Performance Applications III.* New York: Plenum Press, 1983.

Levin, McMurdie, and Hall. *Phase Diagrams for Ceramists.* Columbus: American Ceramic Society, 1956.

Lewis, M. H., ed. Glasses *and Glass-Ceramics.* New York: Chapman and Hall, 1989.

Mangles and Messing, eds. *Advances in Ceramics.* Vol. 9. Columbus: American Ceramic Society, 1983.

Mason, Warren, ed. *Physical Acoustics.* New York: Academic Press, 1965.

McColm, I. J. *Ceramic Science for Materials Technologists.* Glasgow: Leonard Hill, 1983.

McLellan, George, and Shand. *Glass Engineering Handbook.* New York: McGraw-Hill, 1984.

McMillan, P. W. *Glass-Ceramics.* New York: Academic Press, 1979.

Michaels and Chissick, eds. *Asbestos.* New York: John Wiley and Sons, 1979.

Morrell, R. *Handbook of Properties of Technical and Engineering Ceramics: Part 1.* London: Her Majesty's Stationary Office, 1985.

Morrell, R. *Handbook of Properties of Technical and Engineering Ceramics: Part 2.* London: Her Majesty's Stationary Office, 1987.

Moulson and Herbert. *Electroceramics.* London: Chapman and Hall, 1990.

Musikant, S. *Optical Materials.* New York: Mercel Dekker, 1985.

National Materials Advisory Board. *Magnetic Materials.* Publication NMAB-426. Washington: National Academy Press, 1985.

Nickel, Nichols, and Monte. *Mineral Reference Manual.* New York: Van Nostrand Reinhold, 1991.

Norton, F. H. *Elements of Ceramics.* Cambridge: Addison-Wesley Press, 1952.

O'Bannon, L. S. *Dictonary of Ceramic Science and Engineering.* New York: Plenum Press, 1984.

Pampuch, Roman. *Ceramic Materials.* New York: Elsevier, 1976.

Parmelee, C. W. *Ceramic Glazes.* Boston: Cahners Books, 1973.

Paul, A. *Chemistry of Glasses.* New York: Chapman and Hall, 1990.

Pauling, Linus. *College Chemistry.* San Francisco: W. C. Freeman and Company, 1956.

Pauling, Linus. *The Nature of the Chemical Bond.* Ithaca: Cornell University Press, 1960.

Ramachandran, Feldman, and Beaudoin. *Concrete Science.* London: Heyden, 1981.

Ryan, W. *Properties of Ceramic Raw Materials.* New York: Pergamon Press, 1978.

Schwartz, Mel, ed. *Engineering Applications of Ceramic Materials.* Metals Park: American Society for Metals, 1985.

Schwartz, Mel, ed. *Handbook of Structural Ceramics.* New York: McGraw-Hill, 1992.

Searcy, Ragone and Colombo, eds. *Chemical and Mechanical Behavior of Inorganic Materials.* New York: Wiley-Interscience, 1970.

Shugg, W. T. *Handbook of Electrical and Electronic Insulating Materials.* New York: Van Nostrand, 1986.

Singer, F. and S. Singer. *Industrial Ceramics.* London: Chapman and Hall LTD, 1963.

Solomon and Hawthorne. *Chemistry of Pigments and Fillers.* New York: John Wiley and Sons, 1983.

Swamy, R. N., ed. *New Reinforced Concretes.* London: Surry University Press, 1984.

Tallan, N. M., ed. *Electrical Conductivity in Ceramics and Glass.* Volumes A and B. New York: Marcel Dekker, 1974.

The Institution of Metallurgists. *Composite Materials.* London: Iliffe Books Ltd., 1966.

Thompson, D. P., ed. *Engineering Ceramics: Fabrication Science and Technology.* London: The Institute of Materials, 1993.

Tressler, and McNallan, eds. *Ceramic Transactions.* Volume 10. Westerville: The American Ceramic Society Inc, 1990.

U.S. Department of Commerce. "A Competitive Assessment of the United States Advanced Ceramics Industry." #PB84-162288. Washington: U.S. Government Printing Office, 1984.

Ubbelohde, and Lewis. *Graphite and Its Crystal Compounds.* Oxford: The Claredon Press, 1960.

Varshneya, A. K. *Fundamentals of Inorganic Glasses.* New York: Academic Press, 1994.

Wachtman, J. B., ed. "Mechanical and Thermal Properties of Ceramics." *National Bureau of Standards Special Publication 303.* Washington: U.S. Government Printing Office, 1969.

Wachtman, J. B., ed. *Structural Ceramics.* New York: Academic Press, 1989.

Warren, R., ed. *Ceramic-Matrix Composites.* New York: Chapman and Hall, 1992.

Weast, Robert. *CRC Handbook of Chemistry and Physics.* Boca Raton: CRC Press, 1983.

Wills, Roger. *Ceramics: Advances and Opportunities.* Columbus: Battele Technical Inputs, 1981.

Worrall, W. E. *Clays and Ceramic Raw Materials.* New York: Elsevier Applied Science Publishers, 1986.

Zackay, Victor, ed. *High-Strength Materials.* New York: John Wiley and Sons, 1965.

Zarzycki, J. *Glasses and the Vitreous State.* Cambridge: Cambridge University Press, 1991.

INDEX

269

271

DATE DUE

GAYLORD No. 2333   PRINTED IN U.S.A.